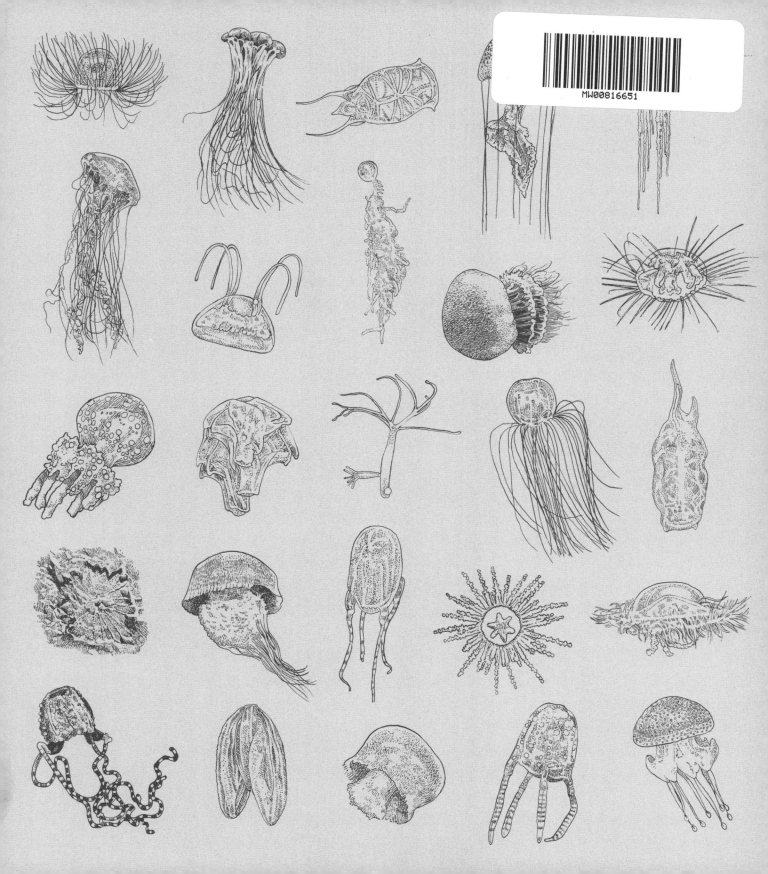

# JELLYFISH

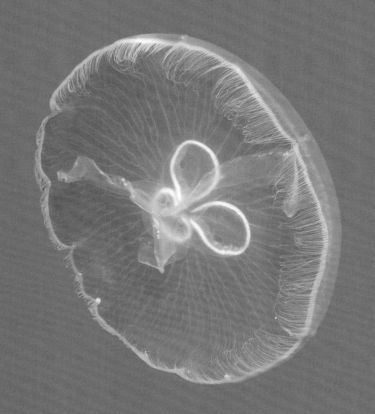

# JELLYFISH

## A NATURAL HISTORY

LISA-ANN GERSHWIN

*To Rod & Cindy —*
*Very glad to meet you! I hope*
*you enjoy getting to know jellies*
*a bit more through these pages!*

*San Pedro 2 Nov 2016*

THE UNIVERSITY OF CHICAGO PRESS

Chicago

**Lisa-ann Gershwin** is director of the Australian Marine Stinger Advisory Services. She was awarded a Fulbright in 1998 for her studies on jellyfish blooms and evolution, and she has discovered over two hundred new species—including at least sixteen types of jellyfish that are highly dangerous, as well as a new species of dolphin. She is the author of *Stung! On Jellyfish Blooms and the Future of the Ocean.*

The University of Chicago Press, Chicago 60637
Text © 2016 Lisa-ann Gershwin
Design and layout © 2016 by The Ivy Press Limited
All rights reserved. Published 2016.
Printed in China

25 24 23 22 21 20 19 18 17 16     1 2 3 4 5

ISBN-13: 978-0-226-28767-6          (cloth)
ISBN-13: 978-0-226-28770-6          (e-book)

DOI: 10.7208/chicago/9780226287706.001.0001

Library of Congress Cataloging-in-Publication Data

Gershwin, Lisa-ann, author.
  Jellyfish : a natural history / Lisa-ann Gershwin.
      pages cm
  Includes index.
  ISBN 978-0-226-28767-6 (cloth : alk. paper) — ISBN 978-0-226-28770-6 (e-book)
1. Jellyfishes. 2.  Natural history.  I. Title.
  QL377.S4G46 2016
  593.53—dc23
                        2015034610

This book was conceived, designed, and produced by
**Ivy Press**
210 High Street, Lewes
East Sussex BN7 2NS
United Kingdom
www.ivypress.co.uk

*Publisher* Susan Kelly
*Creative Director* Michael Whitehead
*Editorial Director* Tom Kitch
*Art Director* James Lawrence
*Commissioning Editor* Kate Shanahan
*Editors* David Price-Goodfellow and Amy K. Hughes
*Design* J.C. Lanaway
*Illustrator* Vivien Martineau
*Species Illustrations* David Anstey
*Picture Researcher* Katie Greenwood
*Editorial Assistant* Jenny Campbell

Cover photograph © Doug Perrine/Nature Picture Library

# CONTENTS

# INTRODUCING
# THE JELLYFISH

Jiggly, flowing, mesmerizing, alien, delicious, stingy, lethal—jellyfish are many things to many people. To the fisherman, they may be a nuisance, sometimes a costly one. To the swimmer, they mean painful and even dangerous stings. To the artist, they may stimulate and inspire. To certain entrepreneurs, they are promising sources of innovation and profit. To the curious, they are endlessly weird and fascinating.

JELLYFISH AS A GROUP hold some astonishing records. The world's most venomous animal is a jellyfish, the Australian Deadly Box Jellyfish (*Chironex fleckeri*, page 50). The largest invertebrate discovered in the twentieth century is a jellyfish, the so-called Black Sea Nettle (*Chrysaora achlyos*, page 114)—though it is practically a toy compared to the lion's mane jellies of the North Atlantic (*Cyanea* spp., page 52), which can reach three meters (ten feet) across the body and drag tentacles nearly 30 meters (100 feet) long. One jellyfish helped scientists win the Nobel Prize (page 198). Another grows ten percent of its body length *per hour* (page 208). And the world's first known case of true biological immortality was discovered in the diminutive and aptly named Immortal Jellyfish (*Turritopsis dohrnii*, page 74).

## The Problem with Jellyfish

We have always been put off jellyfish by their stings, and these creatures have long drifted under the radar scientifically and industrially. But a couple of decades ago, jellyfish started becoming harder to ignore. People began reporting problems, which often interfered with human enterprise, and as the frequency of these reports increased it led to further attention and thus even more reports.

Most of the problems involve jellyfish blooms, or large, concentrated swarms. Jellyfish bloom as a natural part of their life cycle. But increasingly, it appears, some blooms are lasting

far longer than normal, or are covering vastly larger areas, or are considerably denser than usual. It can be challenging to tease apart what is happening in these cases. Are the jellyfish simply misbehaving, or is human activity impinging on their turf and stimulating rambunctious reactions? Regardless, jellyfish are a problem when they threaten lives or livelihoods.

**Below** Jellyfish run the full gamut from beauty to beast, from lovely to lethal. They are united by their gelatinous bodies, drifting lifestyle, simple organ systems, and ability to bloom prolifically. Jellyfish come in an array of shapes and sizes, from those barely larger than a grain of sand to others longer than a blue whale. Some, like *Pyrosoma* (left), are herbivores, eating phytoplankton. Others, like the Flower Hat Jelly *Olindias* (center), are carnivores, eating zooplankton. And some, like the blubbers (right), are both.

## What Is a Jellyfish?

Despite the sensational attention given to jellies in the media, many people are still not sure what, exactly, jellyfish are. They are animals, though they lack recognizable body parts like faces and bones and, in most, a brain and a heart. They are invertebrates, which means they have no vertebral column, or spine, but they belong to different invertebrate groups. Some are in the same category as the corals, sea anemones, and sea fronds; others belong to the lineage that eventually gave rise to humans, and even possess a rudimentary heart and brain but are so primitive evolutionarily that we bear few features in common.

It is also important to keep in mind that not all squishy, drifting, transparent aquatic creatures are jellyfish. Many species of squid and even a few octopus species, for example, are transparent and squishy, and some even drift on the currents. Some fish, particularly the larvae of eels, are transparent and gelatinous (jelly-like). Certain drifting, gelatinous sea cucumbers look more like jellyfish than many jellyfish do. Even the minuscule, spherical Sea Sparkle (*Noctiluca scintillans*), an algae-like organism, could be mistaken for a jellyfish. But none of these creatures are jellyfish. This means the answer to the question "what is a jellyfish?" is as slippery as jellies themselves: Jellyfish are squishy aquatic creatures, some drifting and some transparent, that span different animal groups. But even the experts cannot agree on the group as a whole: some exclude pelagic tunicates (salps and their kin), while others include them, as we do in this book (pages 70–71).

Despite their strangeness, jellyfish, like all animals, must catch food, reproduce, move around, and protect themselves—and they do so without brains, bones, or blood. Jellyfish have been making do for hundreds of millions of years, since before brains, bones, and blood evolved, and their simplicity works for them. They can subsist on a broad variety of food—or even no food at all—and they can reproduce in an exciting variety of ways, with or without a mate. And in their different life stages, they can move or not move, and they get along just fine. It is no wonder they have persisted for so long. They have mastered the art of survival.

The weird and wonderful world of the jellyfish contains many splendid surprises, from their remarkable ability to clone replicates of themselves that are so unlike one another that they have been classified as different creatures, to their tenacious persistence in conditions that most other animals find utterly unlivable, to their delightful shapes and color patterns and mesmerizing movements. The strange otherworldliness of jellyfish makes them simply fascinating.

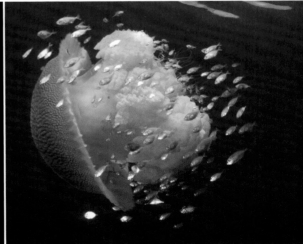

| Group | Examples | Main impacts |
|---|---|---|
| **True Jellyfish** Cnidaria: Scyphozoa: Semaeostomeae | Sea nettles, moon jellies, Purple People Eater | Clogging intake pipes of power plants and ships; stinging |
| **Blubber Jellies** Cnidaria: Scyphozoa: Rhizostomeae | Bazinga, barrel jellies, sea blubbers, cabbage jellies, Sea Tomato | Clogging fishing nets and intake pipes of power plants and ships |
| **Coronate Jellies** Cnidaria: Scyphozoa: Coronatae | Santa's Hat Jelly, Flying Saucer Jellyfish, thimble jellies | Shifting ecology of Norwegian fjords; consume larvae and plankton |
| **Box Jellyfish** Cnidaria: Cubozoa | Box jellies, Irukandjis | Stings very painful and may be lethal or cause severe illness |
| **Water Jellies** Cnidaria: Hydrozoa: Hydroidomedusae | Bell jellies, Nobel Jelly | Predation on and competition with fish eggs, larvae, and plankton |
| **Siphonophores** Cnidaria: Hydrozoa: Siphonophora | Long stingy stringy thingies, Portuguese Man-of-war | Stinging; strong predation on and competition with other species |
| **Comb Jellies** Ctenophora | Sea walnuts, sea gooseberries | Predation on and competition with fish eggs, larvae, and plankton |
| **Salps and Their Kin** Chordata: Tunicata | Salps, pyrosomes, doliolids | Grazing on phytoplankton; compete with other species |

siphonophore    true jellyfish    salp    box jellyfish    comb jelly    coronate jelly    blubber jelly    water jelly

# INTRODUCING
# THE OCEAN

It has been said that our planet should actually be called Ocean rather than Earth, because nearly three-quarters of it is covered in water. This aquatic portion (72 percent of the planet's surface) is the domain of the jellyfish, which are found from pole to pole, from the surface to the depths. Scientists have developed terminology to describe the multitude of different oceanic zones that exist. Some of these terms are used throughout the book, including in the fact panel that accompanies each species account.

IN GENERAL, we divide the ocean into zones guided by either vertical position in the water column or horizontal coverage based on relationship to landforms. Within these zones, numerous habitats characterize different ecosystems and define the flora and fauna that live there. These zones are similar in function to the mountains, deserts, rivers, and lakes that divide up the three-dimensional habitats on land and define the ecospace in which certain organisms can thrive.

## Horizontal Ocean Zones

The "horizontal" ocean zones begin at the coast, where land and water meet. The area of the shoreline that is traversed by the tide each day is called the intertidal zone; only the hardiest organisms generally survive here, due to the extreme temperature, salinity, and wet/dry and motion pressures associated with the incoming and outgoing tide. Jellyfish sometimes become stranded in the intertidal zone

**Right** The Great Barrier Reef in Australia is famous for its corals. The mesmerizingly clear blue water that entices us so, is generally low in nutrients. This type of ecosystem is one of the types of habitat where jellyfish flourish.

## OCEANIC ZONES

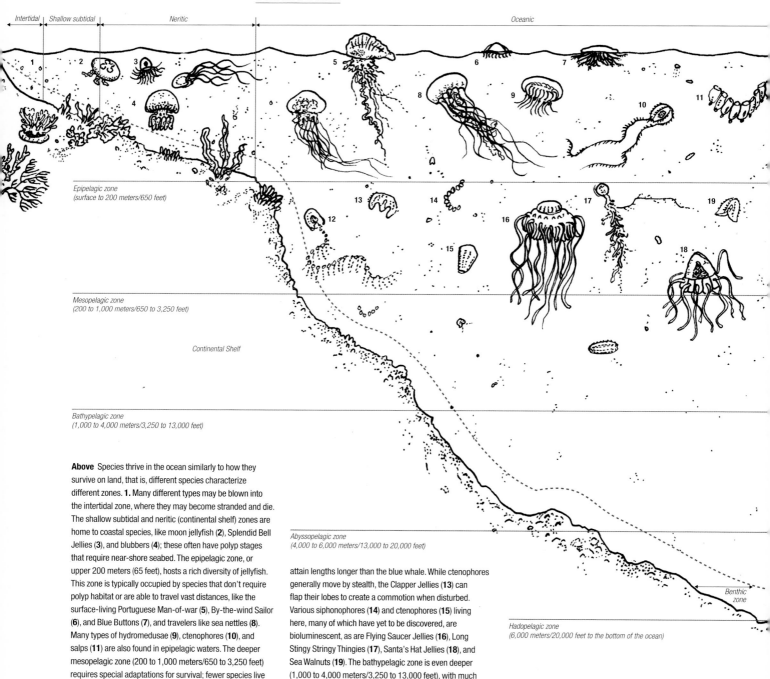

*Intertidal* | *Shallow subtidal* | *Neritic* | *Oceanic*

*Epipelagic zone*
*(surface to 200 meters/650 feet)*

*Mesopelagic zone*
*(200 to 1,000 meters/650 to 3,250 feet)*

*Continental Shelf*

*Bathypelagic zone*
*(1,000 to 4,000 meters/3,250 to 13,000 feet)*

*Abyssopelagic zone*
*(4,000 to 6,000 meters/13,000 to 20,000 feet)*

*Benthic zone*

*Hadopelagic zone*
*(6,000 meters/20,000 feet to the bottom of the ocean)*

**Above** Species thrive in the ocean similarly to how they survive on land, that is, different species characterize different zones. **1.** Many different types may be blown into the intertidal zone, where they may become stranded and die. The shallow subtidal and neritic (continental shelf) zones are home to coastal species, like moon jellyfish (**2**), Splendid Bell Jellies (**3**), and blubbers (**4**); these often have polyp stages that require near-shore seabed. The epipelagic zone, or upper 200 meters (65 feet), hosts a rich diversity of jellyfish. This zone is typically occupied by species that don't require polyp habitat or are able to travel vast distances, like the surface-living Portuguese Man-of-war (**5**), By-the-wind Sailor (**6**), and Blue Buttons (**7**), and travelers like sea nettles (**8**). Many types of hydromedusae (**9**), ctenophores (**10**), and salps (**11**) are also found in epipelagic waters. The deeper mesopelagic zone (200 to 1,000 meters/650 to 3,250 feet) requires special adaptations for survival; fewer species live here but they tend to have fascinating aspects to their biology or ecology. Siphonophores like the Giant Heart Jellies (**12**) can attain lengths longer than the blue whale. While ctenophores generally move by stealth, the Clapper Jellies (**13**) can flap their lobes to create a commotion when disturbed. Various siphonophores (**14**) and ctenophores (**15**) living here, many of which have yet to be discovered, are bioluminescent, as are Flying Saucer Jellies (**16**), Long Stingy Stringy Thingies (**17**), Santa's Hat Jellies (**18**), and Sea Walnuts (**19**). The bathypelagic zone is even deeper (1,000 to 4,000 meters/3,250 to 13,000 feet), with much fewer species still, which are often smaller. Jellyfish do inhabit the deepest oceans but are generally poorly known.

when the tide goes out; more often than not, this is lethal to them. By the time the beach floods again with the incoming tide, they are already too overheated and dehydrated to survive.

Offshore of the intertidal zone is the subtidal zone, which begins at the point along the shoreline that is always underwater, regardless of high or low tide. The subtidal is loosely divided into shallow subtidal, which is within diving depths; neritic, which is the water that lies over continental shelves; and oceanic, or open ocean, which is off the shelf. Jellyfish occupy all these biomes, but different species characteristically occupy different zones.

## Vertical Ocean Zones

The water column—which extends vertically from the surface to the bottom—is called the pelagic realm. The creatures that live here, including many jellyfish, are said to have a pelagic lifestyle. The bottom, or seabed, is the benthic zone. Starfish, clams, and burrowing worms are examples of benthic species. Many benthic species have pelagic larvae, and some pelagic creatures, including many jellyfish, have benthic stages. There are even jellyfish, such as stauromedusae and platyctene ctenophores (page 18), that are primarily benthic rather than pelagic.

Within the pelagic realms, creatures may specialize in a particular lifestyle. Those that drift on the currents are referred to as planktonic (from *planctos*, Greek for "drifter"), whereas those that swim strongly enough to fight a current are called nektonic. Tuna and swordfish are good examples of nekton, whereas most jellyfish are plankton. The term plankton is a catchall grouping for adult and larval organisms at the mercy of the currents.

The pelagic realm is not uniform, and the creatures that live within it vary as well. Working downward from the ocean surface, the epipelagic zone occupies the top layer of water to the depth to which light penetrates sufficiently to allow photosynthesis, about 200 meters (650 feet) down. From there to about 1,000 meters (3,280 feet) deep is the mesopelagic zone—also called the twilight zone—where light reaches only faintly. Mesopelagic organisms often have peculiar

adaptations for coping with small amounts of light, such as bioluminescence, large eyes, or vertical migration behavior.

From 1,000 meters to 4,000 meters (13,000 feet) is the bathypelagic zone; from 4,000 to 6,000 meters (13,000 to 20,000 feet) lies the abyssopelagic zone; and below that the hadopelagic zone drops to the greatest depths. These deeper zones are in permanent darkness, and the organisms that live in them are often blind. Fewer jellies live in these zones than the ones above.

## Habitats

Within the ocean's horizontal and vertical zones are numerous habitats, some of which are commonly inhabited by jellyfish. The dangerous box jellies and Irukandjis (pages 50 and 154), for example, are often encountered along sandy beaches, where they hunt for food. The upside-down jellyfish (*Cassiopea*, page 48) is often found in shallow subtidal lagoons between coral reefs; reefs can also help form eddies in which jellies become entrained. Estuaries—areas where rivers and creeks flow into the ocean—often host vast blooms of jellies. Water flows more slowly in estuaries than it does along open coastlines, and this slower turnover time helps jellies stay inside their boundaries. Industries, ports, and urban areas are often clustered at estuaries, and they may contribute to these blooms and suffer their effects.

Perhaps the strangest habitats of all are at the surface of the water: The neuston zone is the top few inches of the water column, where the surface acts as an uncrossable barrier, concentrating planktonic organisms just below it. The pleuston zone is right at the air-water interface (pages 148–49). Here some organisms live on top of the water, others live by clinging to the underside of the air, and the majority have body parts in both realms. The Portuguese Man-of-war (page 34), with its float in the air and its tentacles in the water, is a familiar example of a pleustonic organism.

Finally, not all jellyfish are found in marine environments. Some species live exclusively in freshwater habitats such as lakes and reservoirs, which lack tides and are generally too shallow to have complex horizontal and vertical zones.

# ABOUT THIS BOOK

This book is divided into five chapters, which cover the primary themes of jellyfish biology and ecology. Organizing the material this way invites us to understand these splendid creatures through features that unite them as well as those that divide. Each chapter examines nine concepts that illustrate each theme and profiles 10 jellyfish species that help bring to life these features.

**Chapter One**

In the first chapter, "Jellyfish Anatomy," the emphasis is on the functional morphology or functional structure of each body part. As we explore the variety of structures used for defense, locomotion, food capture, and so on, we learn how jellies accomplish the basic tasks of eating, moving, and reproducing. We get to know food-capturing structures that range from tentacles bearing piercing, venomous harpoons, to curtains of entangling, sticky threads, to highly modified lures that mimic other creatures. We also look at the different parts of the jellyfish. For example, we look into their eyes, which run the gamut from completely absent, to light-sensing organs that see only brightness and darkness, to complex visual structures with lenses, retinas, and corneas—similar to our own human eyes—that may discern shapes and colors.

**Chapter Two**

"Jellyfish Life History" compares and contrasts the developmental aspects of the different types of jellies, which typically share within their group similar developmental pathways. For example, jellies of one major grouping (the phylum Cnidaria) typically have a complex life cycle with two major stages, polyp (a budlike form) and medusa (the familiar floating bell or umbrella shape). They are literally shape-shifters, assuming different forms at different times of their lives that do not look remotely alike. We also look at some of the ways jellyfish produce medusae: some involve a highly complex segmenting process called strobilation, while others simply bud new medusae, and still others undergo total metamorphosis. In this chapter we also examine such curious transformations as cloning and immortality.

**Chapter Three**

Moving on to "Jellyfish Taxonomy and Evolution," we delve into the extraordinary world of jellies throughout time. We typically view jellies as so completely foreign that we may even wonder whether they are indeed animals or how they came to be so strange. In this chapter we explore how scientists have come to understand jellies' evolutionary history through fossils left behind in ages past and through the relationships revealed by their unusual genetics. We also compare the different species concepts—changes in the ways scientists have considered what a "species" is and how it should be studied—that underlie how we have viewed jellyfish over time.

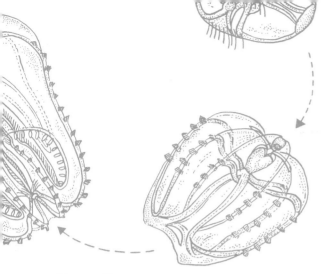

## Chapter Four

In "Jellyfish Ecology," we explore how jellies interact with neighboring species and with the physical features of their environment. The name of the game in the ocean is "eat or be eaten"—and live long enough to reproduce. Over millions of generations jellies have mastered the art of survival, persisting long after most other species have come and gone. Through studying their ecology, we can learn how their relationships with the physical and biological features around them shape their bodies and behaviors. From predation to defense to locomotion to migration, these simple creatures use a variety of ways to survive in a complex world.

## Chapter Five

The final chapter, "Our Relationship with Jellyfish," profiles the main human-caused stressors that are reshuffling our oceans' ecosystems and opening up ecospace for opportunistic species. As their predators and competitors struggle, jellies flourish. In many disturbed areas, jellyfish provide a visible indicator that the ocean is out of balance. Whether they are causing emergency shutdowns of power plants, clogging fishing nets and capsizing fishing trawlers, or creating profound changes in the food chain, jellies are demanding attention like never before. We examine some of the causes—many of them human-generated—of the apparent rise of the jellies.

## Species Accounts

The species accounts that accompany each chapter offer the reader an opportunity to get to know a few of the more interesting jellyfish.

Many species do not have common names and are usually referred to by their scientific names. For a few, the common names in standard use have been confusingly applied to numerous species or are so generic as to be unhelpful (for example, bell jelly, water jelly, box jelly); some of these species have been given revised common names in this book.

One may wonder why so many jellyfish lack common names, but it is simply because jellies have not gotten the widespread attention afforded to commercial species, such as fish, or aroused popular interest or the studied curiosity of naturalists that birds and mammals enjoy.

Where an account features multiple species, the information panel and map at the bottom of the page cover all the species. Distribution data is limited for many species, so the ranges on some maps are estimates.

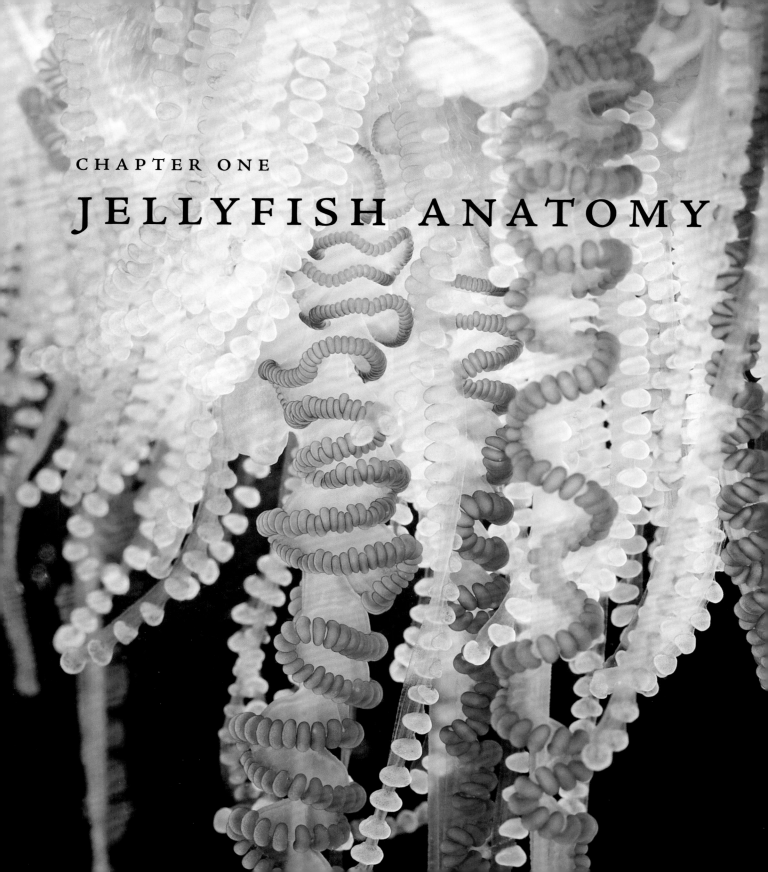

CHAPTER ONE

# JELLYFISH ANATOMY

# INTRODUCTION TO JELLYFISH ANATOMY

Jellyfish, or sea jellies, as they are also known, are simple creatures. Many seem more like plants than animals, although they are animals. Their reproductive capacity can approach rates of exponential growth generally seen only in bacteria, and many of their structural features have no counterpart in more familiar creatures. This chapter provides an overview of the pieces and parts— and their functions—that make up the different groups in the splendidly weird world of jellyfish.

THE GROUP "JELLYFISH" is actually an unnatural collection of creatures from three quite different evolutionary lineages—the medusae and siphonophores, or Medusozoa; the ctenophores; and the salps and their kin— that all just happen to be transparent drifters. Even within these larger groups, smaller subsets often have radically different features. But although they are quite disparate, they share basic elements of their anatomy, which are reflected in their taxonomy, or identification and classification.

All jellyfish have a gelatinous body; a way to catch food, digest it, and get rid of waste; a way to defend themselves; a way to reproduce; and a way to get from point A to point B. But the ways they accomplish these basic life tasks are amazingly and exquisitely varied. These simple creatures are, in many respects, not actually all that simple.

## The Three Lineages of Jellyfish

The most numerous jellyfish—and certainly the ones that are most familiar—are the medusae (singular, *medusa*) in the phylum Cnidaria (pronounced *nye-DARE-ee-uh*, with a silent C), which also contains the corals, sea anemones, and sea fans. Belonging to four classes within the subphylum Medusozoa (Scyphozoa, Cubozoa, Staurozoa, and Hydrozoa), the medusae are generally dome- or disk-shaped but occur in all sorts of unimaginable variations on that theme. They are built on a radially

symmetrical body plan, somewhat resembling a perfectly apportioned pie cut into equal slices, each slice identical to the others. Most species are tetraradial, meaning they have four identical slices, while others have eight (octoradial), and a few have six (hexaradial).

Medusae have the same basic body plan as a coral polyp or sea anemone but upside down; whereas a polyp is basically a stomach opening upward and surrounded with tentacles, medusae are basically the same form facing downward. And, of course, corals and anemones are stuck to the sea bottom, whereas medusae drift in the water.

A subset of the cnidarian class Hydrozoa is a very peculiar offshoot lineage known as the siphonophores (*sigh-FON-uh-fores*). They come in three different body plans consisting of (1) a float and swimming bells, (2) a float but no swimming bells, or (3) swimming bells but no float (see "Siphonophore Life History" on pages 68–69 for more on these forms). Siphonophores are neither predominantly radial nor bilateral, although certain components may be one or the other. They are some of the most difficult of all organisms to identify because they are made up of parts that do not resemble one another or the animal as a whole.

The two other groups of jellyfish are the comb jellies, phylum Ctenophora (*ten-OFF-uh-ruh*), and the salps and their kin, also known as pelagic tunicates, of the phylum Chordata (*kor-DAH-ta*). The creatures of both groups are bilaterally symmetrical—like humans—meaning there is only one way to slice to get equal portions. But in other ways these two groups are quite distinct from each other. Salps may be thought of as a barrel encircled by muscle bands, whereas ctenophores come in a variety of shapes adorned with eight longitudinal rows of large cilia (the "combs" in the name comb jelly).

Despite their different body plans, symmetries, and shapes, and their other various peculiarities, the species we call jellyfish share numerous obvious features. They are pelagic, which means they live in the water column instead of on the seafloor, and they are planktonic, meaning that they drift and, with few exceptions, are unable to fight a current. Their bodies are gelatinous, or jellylike, which helps with buoyancy. And most of them are transparent, which is believed to be a defensive adaptation.

All the structures mentioned in this overview are elaborated on in the sections below. The different groups are treated in more detail in chapter two.

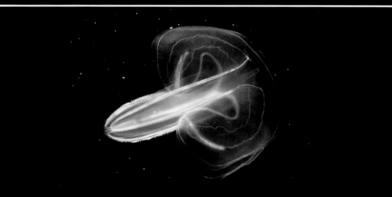

**Right and below** Jellyfish anatomy is often said to be stranger than science fiction. Most medusae, like the stunningly beautiful *Desmonema glaciale* (top right) have a more or less bell-shaped body and fleshy oral arms or long, thread-like tentacles. Oral arms are used for feeding and reproduction, while tentacles are used in feeding and defense. Ctenophores, or comb jellies, like this *Bolinopsis ashleyi* (right) come in the weirdest imaginable shapes: belts, spheres, pockets, walnuts, and many other forms. Salps like this *Cyclosalpa* (below) have a barrel-like body. Even though salps are jelly-like, they are actually more closely related to humans than to other jellyfish.

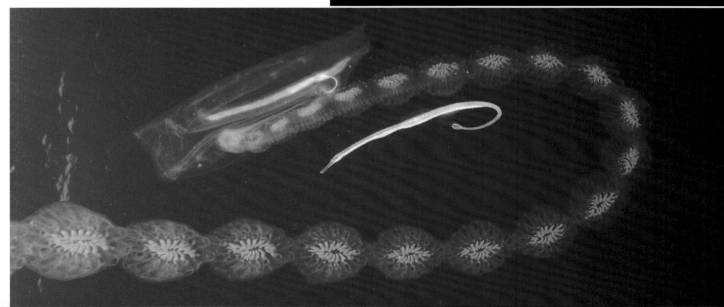

# ANATOMY OF BENTHIC FORMS

While we generally think of jellyfish as drifting organisms, in fact, benthic—or permanently bottom-dwelling—medusae and ctenophores exist, living entirely on the seafloor as adult, sexual forms. In the medusae, these benthic forms are trumpet-shaped creatures known as stauromedusae, which attach themselves to rocks or algae with adhesive glands at the base of a slender, stalk-shaped foot. In the ctenophores, the benthic forms are creeping flatworm-like creatures known as platyctenes.

M OST OF THE stauromedusae and platyctenes (*PLAT-ee-teens*) are unable to swim, but have free-swimming larvae. Even though the adults look totally different and do not drift in the water column, they are still rightfully considered jellyfish because they evolved from jellyfish-like ancestors. While jellyfish in general tend to be poorly known, the stauromedusae and platyctenes are by far the least known of all.

Other benthic forms include the asexual (clonal) stage of the familiar drifting medusae. Hidden away as tiny polyps, they are stuck to rocks, shells, or algae on the seafloor, or the benthic zone. These polyp forms, which include the plantlike hydroids, are treated more thoroughly in our examination of jellyfish life history in chapter two.

## Stauromedusae

Exquisitely beautiful creatures shaped like champagne flutes, stauromedusae usually have eight arms radiating out in a star pattern. Each arm ends with a tuft of short tentacles, and each tentacle in turn has a small ball on the end. These tentacles are packed with stinging cells and are used for food capture and defense. Between the arms, many species have special organs called anchors; the function of the anchors is unknown but possibly sensory. Like other medusae, stauromedusae are radially symmetrical. Most species appear externally to be octoradial (eight-parted),

but internally they are tetraradial (four-parted), as are most of their more traditional drifting medusa counterparts. Stauromedusae come in a dazzling variety of colors and patterns that help camouflage them in the red or green algae among which they are often found living.

One of the most interesting things about stauromedusae is that they are essentially upside down of upside down—that is, whereas normal medusae have all the structural features of polyps but upside down, stauromedusae have flipped right back upward again. So while they appear to be "right side up," this orientation came about late in their evolution. Their ancestors are believed to be normal medusae, whose ancestors are believed to be normal polyps.

## Platyctenes

Platyctenes are another enigmatic form. Like stauromedusae, adults are entirely benthic. They are often encountered by scuba divers, home aquarium enthusiasts, and public aquarium visitors, but rarely recognized for what they are. Resembling flatworms, they are essentially an oval-shaped thin film of tissue that glides over sponges and algae and among the spines of sea urchins. The feature that gives away their true ctenophore nature is the tentacles, resembling those of the related sea gooseberries in bearing numerous lateral filaments arranged in one direction, similar to the barbs along one side of a feather.

STAUROMEDUSA

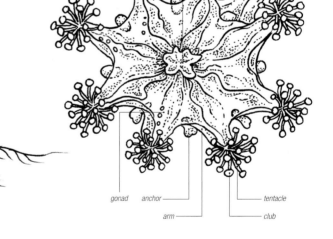

pedal disk

stalk

funnel

gonad

When first hatched, the larval stage of a platyctene looks a lot like a miniature sea gooseberry, and it even drifts in the water column in a similar manner. As it matures, it grows out of its planktonic stage and takes up residence on the sea bottom. One of the biggest challenges a young platyctene encounters is finding a suitable host on which to grow.

Curiously, whereas stauromedusae are essentially medusae turned downside up, so that the mouth faces toward the water's surface, platyctenes are essentially sea gooseberries or sea walnuts flattened in the extreme and turned upside down, with the mouth facing down against the sediment or their host. The body is extremely soft and almost amoeba-like. The upper surface bears numerous ephemeral, intermittently prominent, ciliated papillae, which are believed to serve a respiratory function. Occasionally, two "chimneys" appear, one near each of the two long ends of the body, from which the long feathery tentacles emerge. As with the stauromedusae, platyctenes come in striking colors and patterns, which help them camouflage against their host species.

mouth

gonad   anchor                          tentacle

arm                          club

PLATYCTENE

tentilla

tentacle

chimney

papillae

**Left** Platyctenes resemble flatworms but may be distinguished by their paired feathery tentacles emitting from ephemeral "chimneys," or blunt projections. They glide over their host with their mouth facing down, and lack any trace of comb rows as adults.

**Above** Trumpet-shaped stauromedusae have a similar arrangement to that of coral polyps and sea anemones: the body has a sticky foot at one end for attachment and tentacles encircling the mouth at the other for feeding and defense.

# STINGING CELLS AND STICKY CELLS

The creatures of the two main phyla that contain the jellyfish—Cnidaria and Ctenophora—capture food and defend themselves with the help of two very different types of cellular organelles, or tiny structures made by cells. In the cnidarians these are stinging cells called nematocysts, and in ctenophores these are sticky cells called colloblasts. Both are microscopic.

**Nematocysts: Stinging Cells**

Stinging cells are found in all cnidarians. In fact, the phylum name Cnidaria comes from the Greek word *knidē*, meaning "nettle." This is the primary character that unites such disparate creatures as stony corals, soft corals, sea anemones, sea fans, sea pansies, hydras, medusan jellyfish, and siphonophores.

Stinging cells are wondrous little structures. Each is essentially a double-walled keratinized capsule, with a harpoon coiled up inside and a trapdoor and hair trigger at one end. Because of the hair trigger, the harpoons may discharge with even slight mechanical stimulation. Nematocysts discharge the harpoon with an explosive force of 40,000 Gs, or 40,000 times the force of gravity; their discharge is among the fastest biological processes. Discharge occurs by eversion, or turning inside out, similar to the action when one peels off a rubber glove.

The shaft of the harpoon is hollow, like a hypodermic needle, and often perforated. The venom is contained inside the capsule both on the inside and outside of the harpoon, so that as the harpoon penetrates the skin, it may deliver venom three ways: by hypodermic injection through the tip, along the shaft through the perforations, and by the residue on the outside of the shaft. Strong spines, particularly near the base of the shaft, help anchor the harpoon into prey as it penetrates. The remainder of the shaft may be unarmed or may have three rows of smaller spines spiraling along its length.

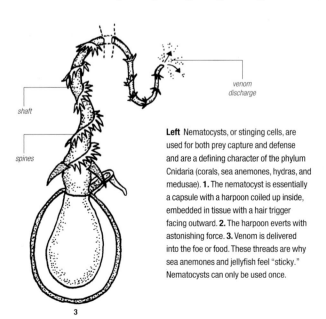

NEMATOCYST

cnidocil
(trigger)

trap door

barb

harpoon

nematocyst

double-walled
keratinized capsule

1

2

3

shaft

spines

venom
discharge

**Left** Nematocysts, or stinging cells, are used for both prey capture and defense and are a defining character of the phylum Cnidaria (corals, sea anemones, hydras, and medusae). **1.** The nematocyst is essentially a capsule with a harpoon coiled up inside, embedded in tissue with a hair trigger facing outward. **2.** The harpoon everts with astonishing force. **3.** Venom is delivered into the foe or food. These threads are why sea anemones and jellyfish feel "sticky." Nematocysts can only be used once.

## Colloblasts: Sticky Cells

The colloblast of the ctenophores is a similarly wondrous structure to the cnidarian nematocyst, though totally different in form and function. Whereas a nematocyst can be thought of as a harpoon full of poison, a colloblast is more like a rope covered in honey. Nematocysts inject; colloblasts ensnare. Colloblasts contain no venom; they are used for food capture but not defense.

The colloblast consists of a bouquet-shaped structure called a collosphere, which bears adhesive granules supported by an axial (central) thread wrapped by a spiral filament. When stimulated, the spiral thread straightens, activating the colloblast, and the granules burst, releasing their glue. Ctenophores with tentacles or lobes have large numbers of colloblasts on these structures; colloblasts are also found on fine tentacles around the lips in some species. One species, *Haeckelia rubra*, lacks colloblasts but co-opts nematocysts from its cnidarian jellyfish prey for its own defense.

## Jellyfish Stings

Because of their venomous nature, nematocysts have been quite well-studied, whereas colloblasts have not. Generally, different types of nematocysts are found in different areas of the jellyfish, such as the bell, tentacles, lips, and stomach. Many dozens of different types of nematocysts have been identified. The structures are typically spherical, ovoid, lemon-shaped, or banana-shaped. The number and the forms of the types present help in species identification in many cnidarian groups. Often they are the only means for distinguishing one species from another, especially following stings, after which fragments of tentacles or nematocysts left in or on the skin may be the only objective evidence available.

Nematocysts can be retrieved from the skin following a sting by a very simple method: adhesive tape is laid sticky-side down on the dry sting area, then pressed to the skin and peeled up. The tape is then put on a microscope slide and examined. This provides a safe, effective, and noninvasive means of identifying the culprit jellyfish.

Treatment of jellyfish stings is largely species specific. For the life-threatening stings of species of box jellies and Irukandjis (see pages 50, 154, and 200), dousing the wound with vinegar will instantly and permanently disable the stinging cells that have not yet discharged, preventing them from injecting any more venom. For other stings, it is just a matter of relieving the pain, which may be accomplished with ice or heat.

COLLOBLAST

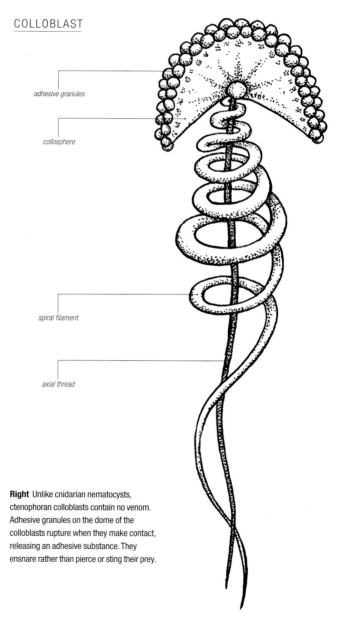

adhesive granules

collosphere

spiral filament

axial thread

**Right** Unlike cnidarian nematocysts, ctenophoran colloblasts contain no venom. Adhesive granules on the dome of the colloblasts rupture when they make contact, releasing an adhesive substance. They ensnare rather than pierce or sting their prey.

# LOCOMOTION STRUCTURES

The majority of jellyfish are essentially passive drifters, at the mercy of the currents. Therefore, even though they may be large, they are nonetheless classified as plankton. Despite their planktonic nature, most possess structures that allow them to change direction, swim up or down, and in some cases, move against weak currents.

**Medusa, Siphonophore, and Salp Locomotion**

The most familiar type of jellyfish locomotion is the pulsation swimming of the medusae. The lesser-known siphonophores and salps employ this mode too. Medusae have well-developed muscles that allow them to contract the bell. Medusae use only half as much energy as any other organism would for the same propulsion, which makes their swimming action one of the most energetically efficient forms of locomotion known. During the contraction, which is the power stroke, the water cupped under the bell is rapidly forced out through an opening narrowed by a thin shelf of tissue called a velum or velarium, creating a thrust of jet propulsion. The counterstroke, when the bell expands back out again, is accomplished through elasticity and "memory" of the bell and therefore requires no energy expenditure.

The cubozoans, or box jellyfish and their kin, have well-developed swimming ability. Some of the larger coastal species, such as *Chironex fleckeri* (page 50), can swim against powerful currents and have been clocked swimming at up to four knots (about five miles per hour). The swimming speed of open-ocean species of the genus *Alatina* has not been measured, but these box jellies have been observed to be very fast and are likely to outpace *Chironex*.

Many types of siphonophores have special structures called swimming bells, or nectophores, used for locomotion. These are basically highly modified medusae that stay stuck to the colony rather than dispersing away and becoming free-floating. Nectophores contract and expand just as normal medusae do, except all the nectophores of the colony must pulsate in a coordinated manner in order to accomplish forward motion.

CONTRACTION AND THRUST MOTION OF A SWIMMING MEDUSA

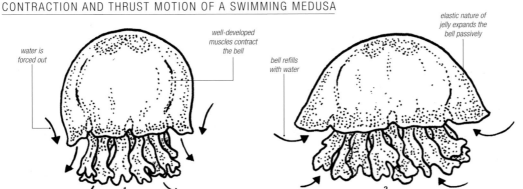

water is forced out

well-developed muscles contract the bell

bell refills with water

elastic nature of jelly expands the bell passively

1

2

**Left** Jet propulsion evolved as an efficient means of jellyfish locomotion long before aerospace engineers dreamed it up. **1.** The bell contracts on the power stroke, pushing water out of the bell cavity, providing thrust. **2.** Then the elasticity of the bell causes it to expand back out to its normal shape, refilling the bell cavity with water. The turbulence created by pulsations disorients the prey, increasing the likelihood of bumping into tentacles.

TRACTOR MOTION OF SWIMMING CTENOPHORE

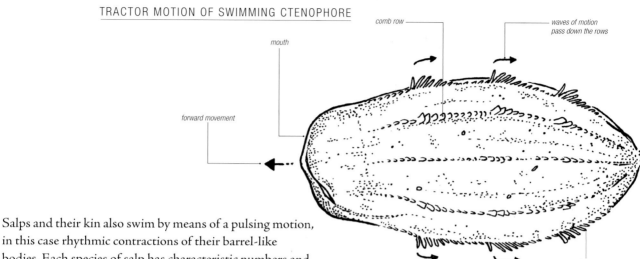

*comb row*

*waves of motion pass down the rows*

*mouth*

*forward movement*

*ctenes*

Salps and their kin also swim by means of a pulsing motion, in this case rhythmic contractions of their barrel-like bodies. Each species of salp has characteristic numbers and patterns of muscle bands encircling the body, like the metal hoops girding a wine barrel. The stimulation of these muscle bands produces the whole-body pulsations that power the animal through the water. Like siphonophores, salps in their aggregate stage need to coordinate their pulsations in order to accomplish forward motion.

**Above** Ctenophores move by stealth, creating very little turbulence. Rows of ciliary plates beat rhythmically, moving the animal by paddle motion on a very small scale. As these rows beat, the plates often catch the light, producing a rainbow effect.

### Ctenophore Locomotion

Only one family of ctenophores is able to pulsate. When touched, species in the family Ocyropsidae (page 40) are able to flap the body violently in a hand-clapping manner, or like a disturbed scallop, and are thus able to rapidly get away from a bothersome stimulus. This wild flapping motion probably also frightens away would-be predators in many cases.

Other species of ctenophores—at least the drifting species—move by means of a sort of tractor motion that involves rhythmic beating of ciliary plates. Recall that ctenophores have eight comb rows or tracks of cilia. These rows are actually composed of dozens to hundreds of clusters of cilia adhered together in small paddles or plates called ctenes (pronounced *teens*). These plates are located more or less along the length of each of the eight rows. A tiny primitive version of a centralized nervous system coordinates the ctenes so that waves of motion pass down the rows

together, accomplishing forward movement with very little vibration. Ctenophores can even go backward, simply by reversing the beating of the ctenes.

### Movement in Benthic Forms

Even the benthic forms of jellies have strategies and structures that help them solve the problem of getting around. Platyctenes lack the comb rows and paddles found in drifting ctenophores, but they are by no means stuck in one place, thanks to their ciliated undersurface. While they normally creep fairly slowly, they can glide at a pretty good clip if necessary. Similarly, although the stauromedusae spend most of their time affixed to rocks and algae, when the need arises, they are perfectly able to move. They simply bend over to one side, discharge some nematocysts to anchor an arm to the ground, release their sticky foot, and somersault away.

# FOOD-CAPTURING STRUCTURES

Jellyfish have a fantastic array of structures used for acquiring food. The most obvious of these structures, of course, are the tentacles. Most people tend to associate tentacles with jellyfish—and vice versa. The tentacles are the features that give jellyfish their creepy or menacing appearance. Tentacles come in a variety of forms. In general, those used to catch larger prey such as fish are thick and heavy, while those used to catch tiny plankton may be as fine as cobwebs (though not always). Regardless of form, the tentacles generally contract or bend to deliver prey to the lips, which then grasp and ingest the food.

### Medusa Tentacles and Oral Arms

Medusan tentacles are typically straight and unbranched, or in rare cases sparsely branched. The tentacles are highly extensible owing to their generally hollow and muscular nature. In some medusae, the nematocysts are arranged along the tentacles in elaborate patterns, such as transverse bands, longitudinal rows, star-shaped clusters or round warts. Lines of nematocysts appear to offer a firmer grip on larger prey than small clusters, which are probably used more for small plankton. In others the tentacle shaft may be without armament, but balls or claws or other structures at the tip may be covered in nematocysts, offering increased stinging surface area. A few species use dangling or glowing structures on the ends of the tentacles to entice prey to come close.

While most types of medusae use tentacles for prey capture, some, such as the rhizostomes, or blubbers, do not have true tentacles and instead concentrate nematocysts on the large amount of surface area of their oral arms, the fleshy structures hanging down from the middle of the body. Other types of medusae, such as the lion's manes and sea nettles, are able to use their oral arms to supplement the food-capturing ability of their tentacles. Lion's manes' arms are complexly folded into a labyrinth of curtains, whereas the sea nettles sport a large, ruffled surface area on their oral arms; in both, the oral arms are loaded with stinging cells. As the animal swims, the oral arms not only catch food but also wipe food off the tentacles and bell margin. Prey are either digested externally or conveyed to the mouth along ciliated grooves.

## Siphonophore Tentacles

Siphonophores, such as the Portuguese Man-of-war (page 34), are similar to the medusae in that they rely on tentacles for food capture. However, because each tentacle is under independent control, the colony must coordinate for maximum efficiency. When working in coordination, the numerous tentacles are set in a long line, forming a sort of curtain of death. When planktonic organisms drift into the zone where these tentacle curtains are deployed, they either drift directly into a tentacle or they startle and bump into one in their escape attempt. Either way, they become prey.

## Ctenophore Tentacles, Lobes, and Teeth

Ctenophores have very sophisticated tentacles compared to those of the medusae and siphonophores. Ctenophore tentacles come in two forms: an obvious, feathered type, which always occurs in pairs that are retractable into sheaths; and the less conspicuous oral tentacles, which surround the mouth in some species. The oral tentacles are simple and straight, while the paired tentacles have hundreds of parallel filaments, called tentilla (singular, *tentillum*), arranged in one plane, like eyelashes. The oral tentacles and tentilla are packed with sticky colloblasts for ensnaring prey.

In addition to the tentacles, many types of ctenophores have large lobes projecting from one end of the body. These are loaded with colloblasts on their inner surfaces. The animal moves toward prey stealthily, as the tractor motion of the ciliary rows makes very little disturbance in the water; once in position, the lobes close around the prey, which, when startled, will bang into one of the colloblast-laden surfaces.

But the most unusual food-catching structures in the jellyfish world would have to be the teeth that occur in some ctenophores (see "Jellies with Bite," page 164). In these species, cilia lining the insides of the mouth have evolved into large, sharp toothlike structures. These teeth are used to bite chunks off their gelatinous prey (mostly other ctenophores).

## Salp Filter Feeding

Salps and kin do not have tentacles, and they accomplish the task of food capture in a completely different way from their fellow jellies. Because they are herbivores, they do not need to sneak up on anything, ensnare it as it drifts by, or kill it. Salps use an internal sticky mucous net to filter tiny phytoplankton (single-celled plants). The net, with openings of graduated sizes, allows them to capture and consume particles spanning four orders of magnitude.

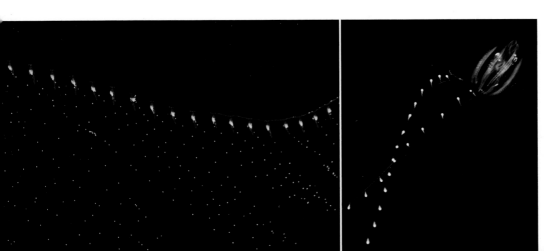

**Left** Jellyfish food capturing structures come in many forms, particularly tentacles. Lion's manes (far left) and sea nettles (center left) can capture prey using their fleshy oral arms, and can digest large items externally, enveloped in the folds of the arms. Most siphonophores (center right) and ctenophores (far right) must deploy their tentacles to ensnare prey. Many siphonophores create a "wall of sting" with numerous tentacles, whereas ctenophores may use their tentacles more like lures or even fishing jigs.

# DIGESTION STRUCTURES

For jellyfish, as for other animals, deriving nutrition from food is fundamental to survival. These creatures have evolved some intriguing structures to accomplish this purpose. The digestive process is essentially divided into three separate phases: ingestion of whole or parts of other animals or plants as food, digestion or extraction and assimilation of nutrients, and excretion of waste.

**The Cnidarian Digestive System**

In hydrozoan medusae and their fellow cnidarians, the mouth functions as both a way to bring food in and a passageway to push waste back out. The structure that bears the mouth is called a manubrium, particularly when it hangs down into the bell; in some species it is very short, or even just a simple hole. In hydrozoan medusae, the stomach, which is called a coelenteron (*sil-EN-ter-on*), is merely an open cavity in the body. In some groups of hydrozoans the stomach is a tube between the mouth and the body; in others it is a broad cavity underlying the mouth in the body itself. Some species also have pores at the base of the tentacles that excrete waste from the digestive system.

Siphonophores are a bit more complex than hydromedusae. Each member of the siphonophore colony has a separate mouth and stomach, but the nutrition is shared throughout the colony via a network of canals that also handle excretion. It is not unusual to witness many mouths from the same colony wrapped around a fish in a bizarre mosaic pattern; however, it may be hard to tell whether they are fighting over the prey or working in cooperation.

**Below** Medusa digestion is simple, involving essentially the intake area, or the mouth; the processing area, or the stomach; the nutrient distribution area, or the radial canals; and the excretion area, typically the mouth again. In scyphozoans, the canals are often quite numerous and branched to supply nutrients to the tissues, whereas in hydrozoans the radial canals are typically few and short. Cubozoans' canals are restricted to their velarium, the flap that extends beyond the body pouches and narrows the bell opening.

## MEDUSA DIGESTION

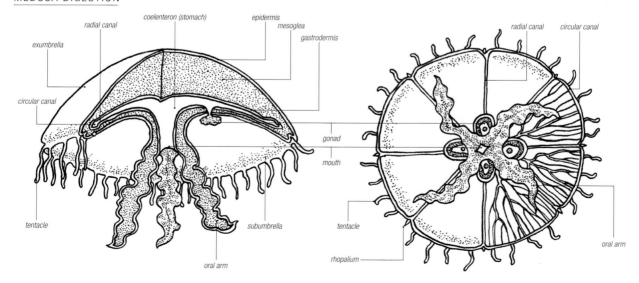

radial canal · coelenteron (stomach) · epidermis · mesoglea · gastrodermis · exumbrella · circular canal · tentacle · subumbrella · oral arm · gonad · mouth · radial canal · circular canal · tentacle · rhopalium · oral arm

**Right** Salps capture food on a continuous stream of mucus and convey it straight to the stomach. Because of the high volume of phytoplankton consumed by salps, the nucleus (gut) is often greenish or brownish.

**Below** Ctenophores are the most ancient creatures to have a through gut, that is, a mouth, intestine, and anus of sorts. Canals branch from the gut to deliver nutrients to the comb rows and gonads.

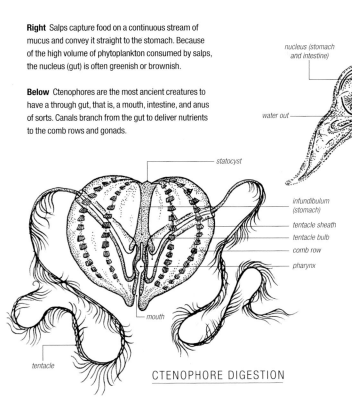

*endostyle*

*nucleus (stomach and intestine)*

*water out*

*water in*

*heart*  *muscle bands*

SALP DIGESTION

*statocyst*

*infundibulum (stomach)*

*tentacle sheath*

*tentacle bulb*

*comb row*

*pharynx*

*mouth*

*tentacle*

CTENOPHORE DIGESTION

Scyphozoan, cubozoan, and staurozoan medusae all have an open cavity, or coelenteron, that is divided into four pouches inside the body. The food and waste both pass through the mouth, as in hydrozoans. A complex series of canals acts as a circulatory system, traveling among the stomach pouches and the bell margin. Most scyphozoans have a single central mouth; however, rhizostomes (blubbers) have hundreds of tiny mouthlets instead, each of which is surrounded by tiny tentacles and opens to the gastric system. These mouthlets are typically arranged in rows on the edges of the oral arms; in some species the arms are complexly branched to increase the surface area, and the mouthlets are distributed all over the food-catching surfaces. Many other scyphozoans use the oral arms in feeding as well. Most just use them to capture extra plankton or to pass food to the mouth, but some are able to digest prey externally on their oral arms if it is too large to fit into the gut; a whole fish takes a couple of weeks to consume when eaten this way.

**The Ctenophore Digestive System**

Ctenophores have more advanced digestive systems than cnidarians. They are the oldest type of animal with a through gut, similar in principle to ours, with separate orifices for food intake and waste output. The food enters through the mouth, pools in a large pharynx, and is then digested in a gut chamber called an infundibulum. The waste is passed through a pair of anal pores, which lie on either side of the centrally located statocyst (balance structure) and ganglion, the part of the animal that controls its nervous actions.

**Salp Digestion**

Salps may be thought of as the vacuum cleaners of the sea. They take water in one end of their barrel-shaped bodies, filter it through a mucous net, and squirt it out the other end. In this way, a single salp can filter hundreds of gallons of water per day. This action provides a feeding current and jet propulsion at the same time. Salps have a through gut that includes an intestine and anus. The mucous net acts as a conveyor belt, constantly funneling food straight into the stomach. Both solid and nitrogenous waste are carried out as water passes through the body, the former excreted as fecal pellets. Because salps are constantly ingesting and digesting a huge volume of food, they produce a large number of heavy fecal pellets; intriguingly, these may act as an important mechanism in sequestering carbon by accelerating its transfer to the deep sea.

# BUOYANCY STRUCTURES

Jellyfish typically spend their whole lives drifting in the water column, and an obvious aspect of their anatomy concerns their need to stay afloat. While they sometimes manage their positioning in the water through behaviors such as active swimming or postures that reduce the sinking rate, buoyancy is more often accomplished through structures that have evolved over the eons. These include mesoglea (the jelly of jellyfish), gas-filled floats, and spikes and other drag-producing appendages.

### Jellyfish Jelly

All jellyfish have "jelly" to some extent, but it is most prevalent in medusae. In cnidarians, including corals and sea anemones, the mesoglea is the jelly that occurs between the ectoderm (outer skin) and endoderm (inner skin). In higher animals a mesoderm (middle skin) occurs between these layers, but in jellyfish, this area is usually just filled with jelly. This lack of a mesoderm is why cnidarians do not have a brain—the brain develops from mesodermal tissues. Salps, however, possess all three embryonic tissue layers, and in fact have a rudimentary central nervous system including a small dorsal ganglion, or brain-like node, complete with a primitive eye.

The jelly gives the body elasticity and the three-dimensional structure that holds the shapes we recognize as different species. The jelly is made of a collagen matrix filled with lots of seawater—about 96 percent. This means that most jellyfish are neither inherently buoyant nor inherently heavy, they are simply about the same density as seawater. The other 4 percent is their own cells, which make up their skin and reproductive organs, so they do not need to exert much effort to stay up in the water column.

Some groups, such as the blubbers and sea nettles, have incredibly thick layers of mesoglea, the mass of which overwhelms their own cellular density, providing buoyancy. In addition to keeping the animal the same density as seawater, and therefore less likely to sink, the large size of the mesogleal mass also helps keep it aloft by slowing its sinking rate.

Many medusa species and some ctenophores and salps have gelatinous spikes and other projections. These appendages probably do not help with defense, because they are not sharp, but they may help slow their owner's sinking rate by adding drag.

An intriguing buoyancy adaptation found in some species of hydromedusae is a large bulbous section of mesoglea on top of the bell, resembling a turret on a mosque or the head on a snowman. Observations that this additional section may vary in size mystified scientists for many years as to its possible function. It was subsequently discovered that this extra mesoglea is used by the medusa to concentrate heavy or light ions in order to facilitate vertical migration. This unusual phenomenon is discussed in greater depth in "Migration" (pages 152–53).

### Gas-filled Floats

By far the most successful flotation devices found among the jellyfish are the gas-filled floats that occur in some siphonophores. Pneumatophores, to give these floats their proper name, come in many shapes and sizes, but all operate similarly: they hold some mixture of gas that the organism can regulate in order to control its position on or in the water. This gas is often just air, but it may be other mixtures. A few species concentrate carbon monoxide, which presents an interesting scientific conundrum, because this gas is highly toxic to living cells.

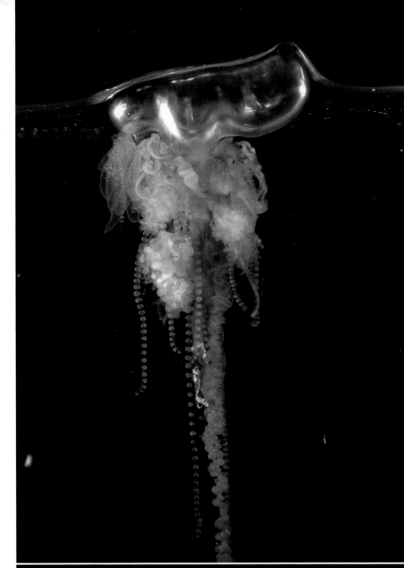

The most familiar example of a gas-filled float is that of the Portuguese Man-of-war (page 34). In this species, the large blue bladder or bubble holds so much air that the float stays above the water, suspending the colony from the surface. Two of the three orders of siphonophores have gas-filled floats; however, most species are found in the deep sea rather than at the air-water interface (i.e., the surface). In some groups, the float has a pore through which the colony can "burp" to release air in order to dive.

Sometimes species with gas bladders swarm in such vast numbers that they create false sonar readings. Military personnel, anglers, and scientists have reported huge shoals of pneumatophore-bearing siphonophores causing a "false bottom" reading on acoustic instrumentation. Studies are now underway to determine whether these shoals may also result in acoustic overestimates of fish stocks.

Siphonophores without gas bladders often possess oil-filled glands, which likely help with buoyancy. The oil in these long, baton-shaped structures may be golden, red, or greenish, depending on the prey it comes from.

**Above right** A hazard of having so much buoyancy is that the float stays above water and can become dry and brittle. *Physalia* uses muscular contractions that briefly pull the float under to solve this problem.

**Right** Thick jelly gives medusae buoyancy, and has other advantages too. For example, it may make its owners appear larger to predators, protect vital organs, and in some, may be used to store oxygen.

# REPRODUCTIVE STRUCTURES

All jellyfish have female or male reproductive structures—ovaries or testes—which produce eggs or sperm, respectively. Some have both. However, these reproductive structures, or gonads, have very different forms and locations in the various jellyfish groups. Some types of jellyfish can even be sexed with the naked eye.

**Cnidarian Reproduction**

Hydrozoan medusae have the simplest gonads of all; they are not proper organs but merely accumulations of sex cells that ripen in particular sites, usually epidermal folds located on the undersurface of the body wall. In species in which the stomach projects down from the body, the gonads are typically located along the sides of the stomach wall. Hydromedusae are usually dioecious (*dye-EE-shus*); that is, the gonads on each medusa are either male or female but rarely both. Typically, the eggs and sperm are just shed into the water. A few hydromedusae, however, will brood their fertilized eggs and the developing larvae until they are ready to swim away.

Most siphonophores, in contrast, are hermaphroditic (both male and female), or monoecious (*mon-EE-shus*). In some cases in the animal kingdom, hermaphrodites can self-fertilize; in many cases this is not possible because maturity of the sexes takes place at different times within each organism. Reproduction in siphonophores is poorly studied and unknown for most species.

Scyphozoans, which include the sea nettles, blubbers, and lion's manes, generally have separate sexes, although there are rare exceptions. Unlike those of the hydrozoans, scyphozoan gonads are located internally. They are separated into distinct tightly coiled organs at the midlines of the four main radial segments of the animal. Visual identification of the sexes is usually easy and can often be made in the field: female gonads typically have a granular texture and diffuse pale brownish,

pinkish, or yellowish appearance, whereas male gonads tend to be creamy-textured and crisply defined, appearing either dark purple (out of water) or bright white (in water). Curiously, while each scyphozoan medusa is a separate sex, research has demonstrated that the asexually produced clone mates from a single polyp may be of either sex; thus, they are clonally hermaphroditic. Many species of scyphozoans have special brood chambers internally or on their arms in which their larvae develop.

## CUBOZOAN GONADS

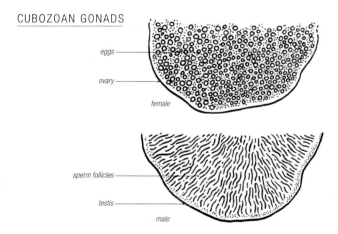

eggs
ovary
female

sperm follicles
testis
male

**Above** Cubozoan reproductive structures (gonads) are thin, leaf-like sheets of tissue. The sexes are easy to tell apart with a hand lens or dissecting scope: females' ovaries have lots of small round eggs, and males' testes look more like a fingerprint.

**Right** Ctenophore sex organs occur together in the same animal, making them hermaphrodites. The ovaries and testes typically occur along the comb rows, with one side being male and the other female. Some species are even able to fertilize themselves.

SCYPHOZOAN GONADS

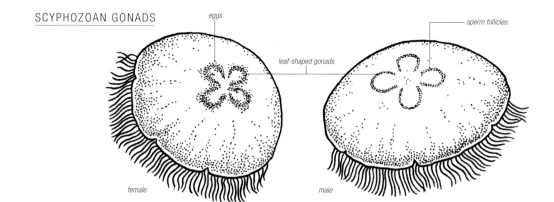

eggs

sperm follicles

leaf-shaped gonads

female

male

**Right** Scyphozoan gonads may be a tightly wound spiral filament or globules in small pouches. Here, *Aurelia* is easily identified by its horseshoe-shaped gonads. In females the gonads appear granular with an indistinct outline, whereas in males they are smoother with distinct outlines.

Cubozoans are always separate sexes. The four gonads are leaf-shaped and attached vertically along the midline in each corner of the bell; thus each corner has two gonad halves projecting laterally toward those on adjacent corners. The female gonads are easy to identify because of the large number of spherical eggs; the male gonads are also fairly simple to distinguish because the sperm follicles are arranged in a fingerprint-like pattern on the broad leaves. Most cubozoans broadcast their sperms and eggs into the water to be fertilized; however, a few species fertilize internally and brood their larvae in special embryo strands that are essentially "laid" among algae.

CTENOPHORE GONADS

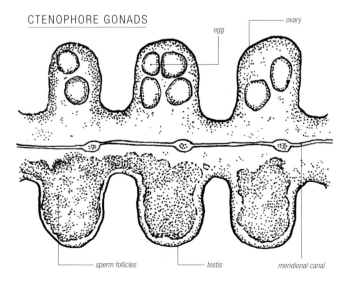

ovary

egg

sperm follicles

testis

meridional canal

Most jellyfish do not mate in the sense that we would recognize. In general, males have neither the equipment nor ability to insert sperm into females: males release sperm threads, and females ingest these to fertilize their eggs. However, one notable exception is the cubozoan Herpes Jellyfish (*Copula sivickisi*, page 88). The male stores sperm in a special vesicle, or internal container, until needed. At the right time, the male grasps a female and deposits a sperm packet on her tentacle; she then ingests it to fertilize her eggs internally.

**Hermaphrodites: Ctenophore and Salp Reproduction**
Almost all ctenophores are hermaphrodites, with both male and female gonads occurring in the same individual simultaneously. The gonads are located beneath the comb rows. In some species, the ovaries occur on one side and the testes are found on the other. Some are able to self-fertilize. Platyctenes are an exception: they are protandrous, or first male and then female.

Salps, too, are hermaphrodites. They have a complex life cycle with alternating generations of sexual and asexual phases (discussed in more detail in "Life History of Salps and Their Kin," pages 70–71). Here we concern ourselves with the reproductive organs of the aggregate sexual phase, in which individuals are stuck together forming chains. Salps start out as female. Larger, older males fertilize the females. An embryo grows inside the female, nurtured through a placenta, and as it matures it triggers the female to become male.

# SENSORY STRUCTURES

The sensory structures of jellyfish run the gamut from negligible to gloriously complex. The structures on which they rely most in their daily lives include a nervous system, which regulates their pulsations and sends various signals around the body; balance structures, which help them maintain their orientation in the water column; and light-sensing structures (rudimentary to complex eyes), which help them gather information about the world around them. There are also indications that at least some jellyfish may be able to sense vibrations and color, and possibly even subtle temperature or chemical changes.

**Cnidarian Sensory Structures**

Hydrozoans (hydromedusae and siphonophores) have a diffuse weblike nerve net rather than a centralized nervous system, and no perceptible nerve nodes, or ganglia (singular, *ganglion*). Siphonophores appear to lack any sort of balance-sensing apparatus or light-sensing structures. Hydromedusae, in contrast, often have both. Their balance structures, called statocysts, operate comparably to our own inner ear. Each statocyst contains numerous tiny granules inside a closed nerve-lined sac, and as the granules move inside the sac, different nerves are stimulated, sending information about the animal's orientation. Many hydrozoan medusae also have red, black, or brown ocelli (singular, *ocellus*), light-sensitive structures that cannot "see" per se but can tell light from dark. When present, the statocysts are located along the rim of the bell between the tentacles, whereas the ocelli are typically found at the bases of tentacles.

Scyphozoans (sea nettles, blubbers, and lion's manes) are a bit more sophisticated or evolved compared to hydrozoans when it comes to their sensory apparatus. They have a nerve net as the hydrozoans do, but they also have eight or more nerve nodes, called rhopalia (singular, *rhopalium*), which control the pulsations of the bell and contain the statocysts and ocelli. So even though the structures and functions are similar, in scyphozoans they are grouped into specific organs.

In cubozoans, or box jellyfish and their kin, instead of the diffuse nerve net, everything is controlled by four rhopalia, which are located inside tiny cavities in the body wall, one on each of the four flat sides, down near the bell margin. These rhopalia are connected by strong nerve cords. In contrast to the numerous tiny granules in the hydrozoan and scyphozoan statocysts, in cubozoans there is only one comparatively large granule, typically called a statolith (Greek for "balance stone"). The cubozoan statoliths are made of gypsum, and the animal adds a thin layer every day, in a manner similar to the way trees add new rings each year. These structures allow scientists to determine the age of an animal. A statolith sits at the downward end of each rhopalium, acting as a counterweight to keep the rhopalium always right side up.

Cubozoans also have incredibly complex visual structures on their rhopalia. Each rhopalium possesses six eyes arranged in three unequal vertical rows. The two side pairs are light-sensing ocelli similar in function but more complex in structure than the ocelli of hydrozoans and scyphozoans. The middle two have lenses, retinas, and corneas and may be able to form images, as human eyes do.

**Sensory Structures of Ctenophores and Salps**

Ctenophores have a semi-centralized nervous system; they do not have a head exactly, but they do have a single nerve point that controls the rest of the body. This is located at the end of

## SIDE VIEW OF A CUBOZOAN RHOPALIUM

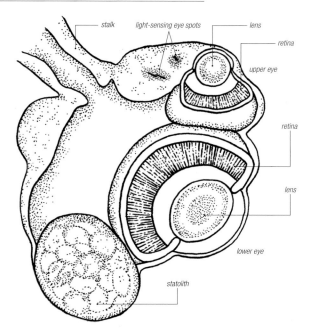

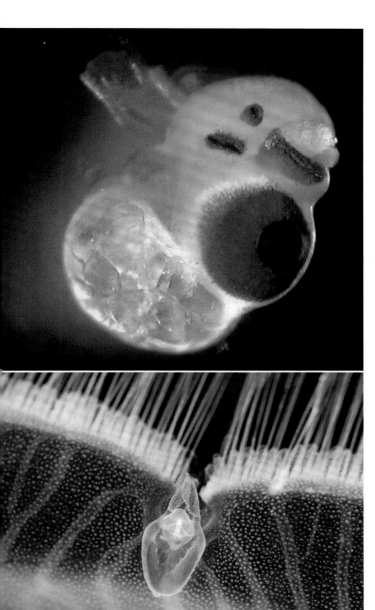

**Left** Scyphozoans have eight tiny finger-like organs called rhopalia that contain a primitive ocellus and a microscopic cluster of granules used for balance. In most species the rhopalia are protected by flap-like growths of the bell margin.

**Above left and above** Cubozoan sensory structures give their owners the ability to see, hunt, court, navigate, and attract some objects or avoid others. Each animal has eight image-forming eyes and up to 16 light-sensing eyes and four balance stones.

the body away from the mouth and contains a statocyst, also used for balance as in the medusae. Ctenophores appear to lack any sort of light-sensing structures.

Salps and kin have a rudimentary central nervous system governed by a tiny brain-like structure called a dorsal ganglion. It is located inside the body near the front end of the animal and it is equipped with a light-sensing ocellus. A balance-sensing structure is present in other types of tunicates, but not salps. Of particular interest from an evolutionary perspective, salp larvae possess a rodlike notochord and nerve cord, which may be thought of as the embryological precursor to the vertebral column and spinal cord, respectively. In larval vertebrates the notochord helps in development and positioning of the spinal cord, but in salps it is lost during maturation.

# PORTUGUESE MAN-OF-WAR AND BLUE BOTTLE

THE MERE MENTION of the Portuguese Man-of-war strikes an ominous tone. It sounds menacing, deadly, foreboding. And yes, it is all those things. And it stings—badly. The Man-of-war's dozens of tentacles each carry enough venom to instantly paralyze a whole school of fish, and entanglement in the tentacles can mean a rapid death, even for a healthy person.

## Fatal Attraction

But the Portuguese Man-of-war (*Physalia physalis*), and its smaller, seemingly less ferocious relative the Blue Bottle (*Physalia utriculus*), are as fascinating as they are dangerous. Their most noticeable feature is the exquisitely blue bladder, or float. When the man-of-war is drifting, the float remains above the water, suspending the tentacles as fishing lures.

During the warmer months these creatures are blown ashore in vast armadas in tropical and subtropical regions. Each "individual" is actually a colony, and most in any given stranding are either right- or left-"handed" in terms of how the crest of the float, or sail, aligns with the rest of the body. Out in the middle of the open ocean where *Physalia* lives and breeds most of the time, the forms are thought to be more or less equally mixed. When a breeze stirs up and pushes the men-of-war toward land, only those with the sail facing the correct way for catching that breeze will be engaged. If the breeze is sustained, they will be pushed until they become stranded on land, where most will perish. Scientists theorize that having these right- and left-oriented forms offers a survival advantage, as part of the population is left behind to keep the species alive.

**Scientific name** *Physalia physalis* and *Physalia utriculus*

**Phylogeny** PHYLUM Cnidaria / CLASS Hydrozoa / ORDER Siphonophora / SUBORDER Cystonectae

**Notable anatomy** large gas-filled bladder (float) with many tentacles

**Position in water column** floats at the air-water interface in open ocean; occasionally blown ashore

**Size** *P. physalis* bladder to 30 cm (12 in.) long; *P. utriculus* bladder less than 10 cm (4 in.) long, more typically less than 5 cm (2 in.) long

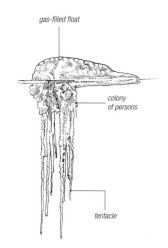

gas-filled float

colony of persons

tentacle

*Distribution*

# PRISM JELLIES

Like other siphonophores, each "individual" of *Vogtia* or *Hippopodius* is actually a colony. They are about the size of a marble, but their "body" is actually a cluster of prisms, like a stack of coins, held together by articulating shapes. The prisms are semi-circular like a horseshoe (*Hippopodius*) or roughly pentagonal, or five-sided (*Vogtia*), and each is turned about 45° relative to those on either side. So they look like balls with lots of bulges, ridges, or corners. The roughly spherical shape is thought to help with buoyancy. One species, *Vogtia spinosa*, also has semi-hard gelatinous spikes—like tiny teeth—covering its outer surfaces; these may slow the rate of sinking by adding extra drag, and may also be useful for defense.

Members of the colony are arranged on the stem, which is neatly tucked away most of the time, withdrawn into a cavity between the protective prisms. However, when hungry, these species unfurl tentacles through the base of the prisms. The tentacles are used for catching prey like copepods, larvae, and other small plankton.

Crystal clear in life, *Vogtia* and *Hippopodius* curiously go opaque white when they die. They are also bioluminescent—like many types of jellyfish—and will flash a bright blue warning light if disturbed.

## Fragile Colonies

*Vogtia* and *Hippopodius* are found regularly in the mid-waters of the oceans and seas of the world, though rarely in high abundance. They are among the most delicate of all jellies, typically disarticulating on contact with a net: scientists consider it a real prize to see an intact colony.

**Scientific name** *Vogtia* spp. and *Hippopodius hippopus*

**Phylogeny** PHYLUM Cnidaria / CLASS Hydrozoa / ORDER Siphonophora / SUBORDER Calycophorae

**Notable anatomy** marble-shaped colony composed of numerous flattened pentagonal or horseshoe-shaped prisms

**Position in water column** epipelagic to mesopelagic, in open ocean

**Size** less than 2.5 cm (1 in.) in diameter

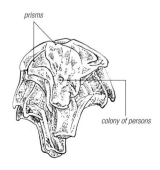

prisms

colony of persons

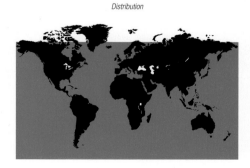

*Distribution*

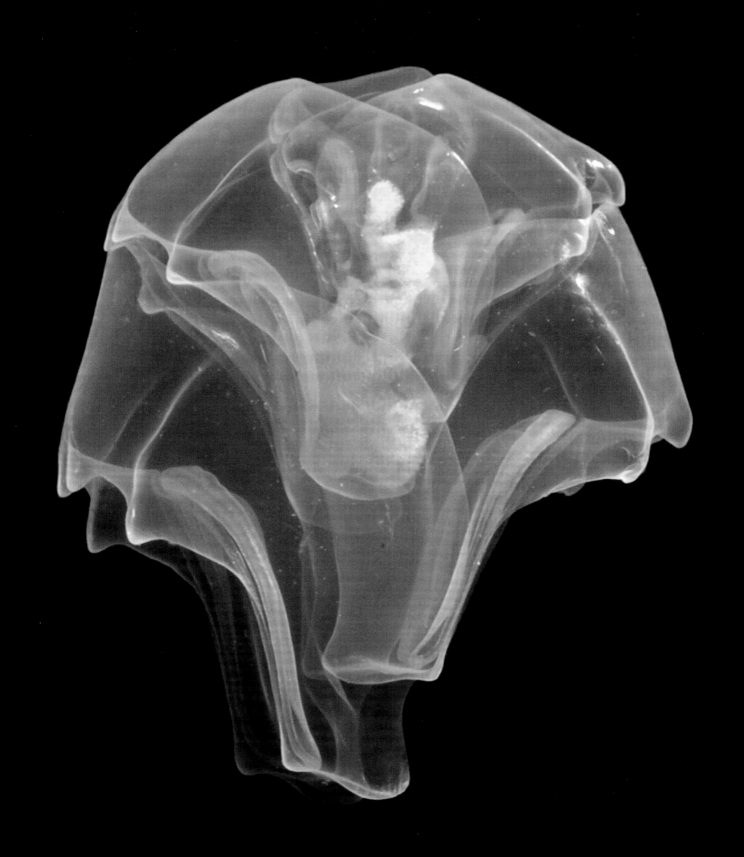

# GIANT HEART JELLY

ONE OF THE MOST IMPRESSIVE siphonophores is the Giant Heart Jelly (*Praya dubia*). This jellyfish can grow to about 50 meters (165 feet) long, which makes it longer than a blue whale. *Praya* colonies consist of two opposing large, semi-hollow gelatinous masses—resembling two tall glass jars side by side, or two sides of a giant heart—with a cylindrical stem about the width of a broomstick protruding from the bottom.

## Persons and Protectors

Members of the colony are arranged in repeating groups along the stem, each containing individuals—known as "persons"—for food collection, digestion, defense, and reproduction. These colony members are protected by thick, solid gelatinous masses called bracts. The bracts are completely transparent and colorless, while the persons they protect typically alternate with yellow and red—warning colors! The tentacles between the bracts can deliver a painful sting, capable of instantly paralyzing their plankton prey or bringing tears to the eyes of human victims.

Giant Heart Jelly is a highly active swimmer, thanks to the two large masses—the nectophores, or swimming bells—which act as powerful jets, their rapid pulsations giving the entire colony its propulsion. When hungry, *Praya* will swim in a three-dimensional spiral pattern and then stop and deploy hundreds of long, transparent tentacles along the length of the colony. This feeding behavior creates an impenetrable curtain of death for plankton and fish. Food caught anywhere along the stem is shared by all the members of the colony.

*Praya* lives mostly in the lower epipelagic to mesopelagic zones, where light penetrates only dimly, if at all. It attracts prey with blue bioluminescence.

**Scientific name** *Praya dubia*

**Phylogeny** PHYLUM Cnidaria / CLASS Hydrozoa / ORDER Siphonophora / SUBORDER Calycophorae

**Notable anatomy** two large, side-by-side columnar nectophores with a thick stem

**Position in water column** epipelagic to mesopelagic, in open ocean

**Size** up to 50 m (165 ft.) long

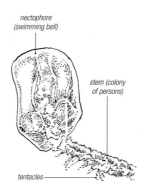

nectophore
(swimming bell)

stem (colony
of persons)

tentacles

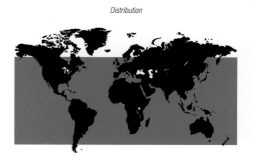

Distribution

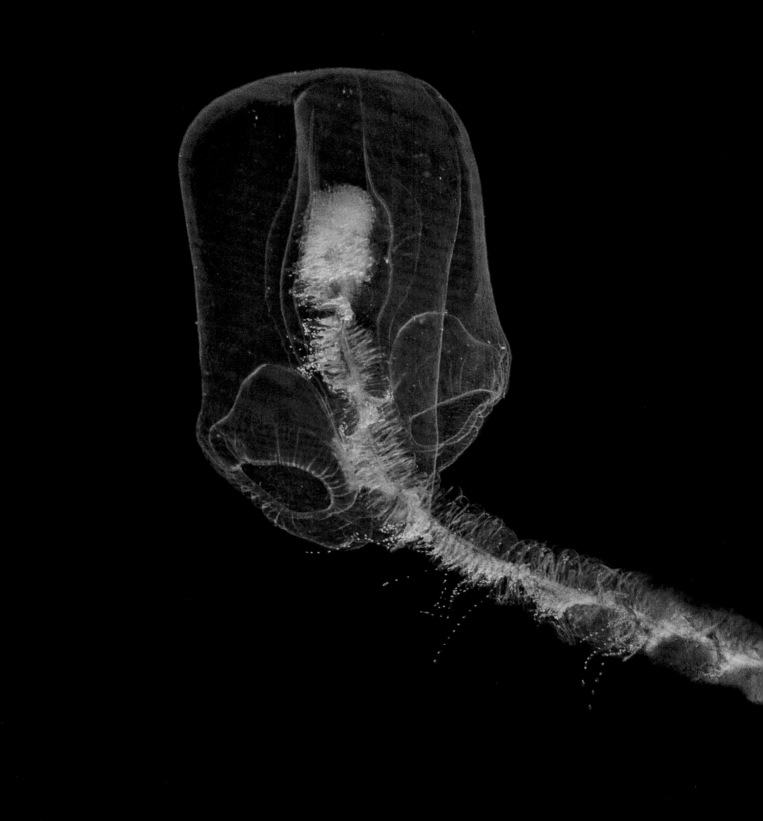

# CLAPPER JELLY

WHILE ALL CTENOPHORES may be legitimately considered quite strange, one of the weirdest and most wonderful is *Ocyropsis*. For one thing, there's how it looks. *Ocyropsis* is shaped like praying hands connected at the palms. When viewed laterally, the body itself is fairly small, surrounded by a pair of enormous opposing lobes. But when viewed from the oral side—that is, looking straight into the mouth—*Ocyropsis* looks like a cross between the Batman logo and what could be a Klingon attack vessel. The lobes protect four rabbit ear-shaped appendages (called auricles) of unknown function.

Another thing that makes *Ocyropsis* so strange is its escape response. Ctenophores normally move forward by means of rhythmically beating ciliary paddles. Many can reverse the motion of their paddles to simply back out of trouble. But *Ocyropsis* gets itself into a flap—literally.

It claps its lobes together in a wild flapping motion to quickly jet away from threats. This also creates great turbulence, no doubt leaving its predators rather surprised.

Like most other ctenophores, species of *Ocyropsis* are hermaphrodites, that is, individuals are both male and female (see page 31). The reproductive organs (gonads) are arranged in pouches along canals underneath the eight comb rows: testes on one side and ovaries on the other.

## Light Defense
Clapper jellies are also bioluminescent, producing their own light. Bright blue-green light primarily flashes along the comb rows. It is thought that bioluminescence is used in this way to startle would-be predators, and possibly also to deter parasites like amphiphods (small crustaceans).

**Scientific name** *Ocyropsis spp.*

**Phylogeny** PHYLUM Ctenophora / CLASS Tentaculata / ORDER Lobata

**Notable anatomy** small body with two enormous lobes

**Position in water column** epipelagic to mesopelagic

**Size** body to about 6 cm (2.35 in.) long

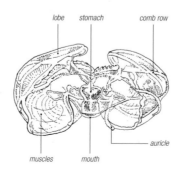

lobe    stomach    comb row

muscles    mouth    auricle

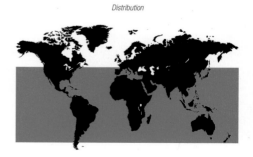

Distribution

# FIRE BODIES

Pyrosomes (Greek for "fire body") are long and tubular sea creatures that are pinkish in color, open at one end, and pointy at the other. They are usually a few inches to a foot or two long. Each "individual" is actually a colony and they are related to sea squirts (tunicates). The colony members are embedded crossways through a semirigid gelatinous matrix, so that their intake valves siphon water from outside and exhale water to the communal chamber inside; this creates a steady flow that jet propels the colony through the sea.

The water flow also brings a rich supply of phytoplankton food, which is sorted through mucous windows and conveyed to the mouth. Pyrosomes are not venomous, and they have no means of defense.

Pyrosomes are among the most gloriously bioluminescent creatures on the planet. When a colony is touched, a wave of blue light flashes outward in a manner reminiscent of the ripples on a pond from a thrown stone. Each zooid is stimulated by the light in turn, and reacts by flashing its own light—and so the wave of light propagates around the colony. As it comes back around, the original zooids can be restimulated, flashing again.

## Night Swimming

Several species are common in the tropical and subtropical seas of the world. One species, *Pyrostremma spinosum*, grows to more than 20 meters (65 feet) long; its colonies form tubes big enough for a person to swim into. Divers encountering them in the Tasman Sea between Australia and New Zealand wait until dark (so the bioluminescence is visible), then position themselves in the opening of the tube, and gently poke it to make it light up.

**Scientific name** *Pyrosoma* spp., *Pyrosomella* spp., and *Pyrostremma* spp.

**Phylogeny** PHYLUM Chordata / SUBPHYLUM Tunicata / CLASS Thaliacea / ORDER Pyrosomatida

**Notable anatomy** long, hollow gelatinous tube with zooids embedded in the wall

**Position in water column** epipelagic, in open ocean and neritic zone

**Size** can be more than 20 m (65 ft.) long, more typically less than 1 m (3 ft. 3 in.) long

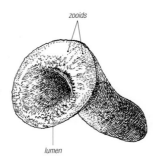

zooids

lumen

Distribution

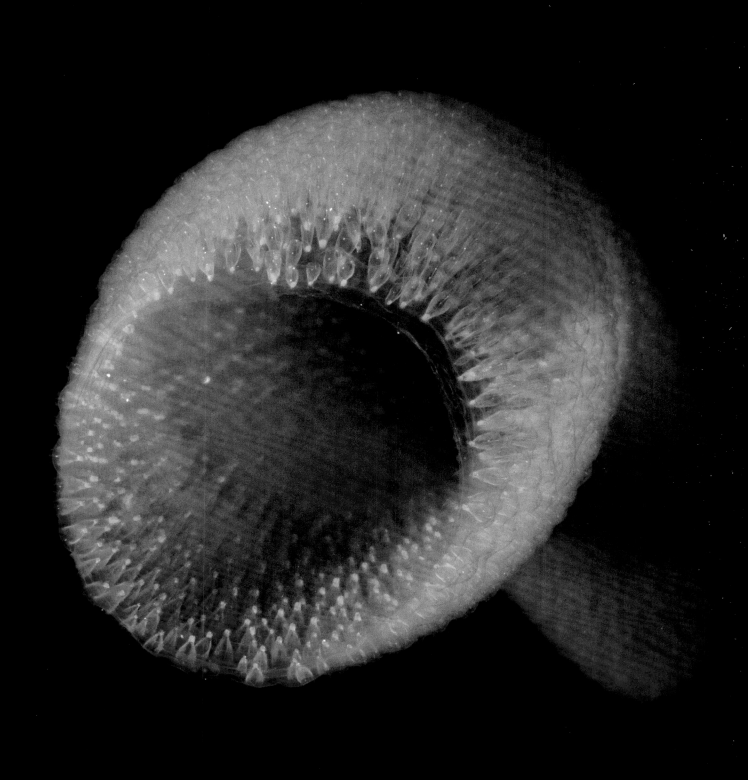

# SEA GOOSEBERRY

THE SEA GOOSEBERRIES, which comprise several families in the order Cydippida, are familiar worldwide to most people who have gone beachcombing after a storm, or gone diving or snorkeling in the summertime, when the water is filled with gelatinous creatures. The completely transparent grape-shaped body bearing eight rainbow-flashing ciliary bands and two long, feathery tentacles is unmistakable. Like most other ctenophores, or comb jellies, sea gooseberries have no ability to sting for defense or to pulsate for locomotion. They swim by coordinating their cilia, employing them as hundreds of tiny oars.

### Gone Fishing

Every now and then, sea gooseberries slow their swimming, unfurl their tentacles, and with a couple of small turns and twists, set them to fish for food. The two tentacles have hundreds of fine side branches oriented all in one direction;

these branches are armed with thousands of microscopic sticky cells. Essentially, the animal hangs two curtains of the marine equivalent of flypaper, and then it sits and waits for food to brush by.

### Somersault Feeding

The sea gooseberry body has two internal cavities into which the tentacles retract and from which they expand. These sheaths run obliquely through the body in such a way that the tentacles face backward, away from the mouth, presenting challenges in the way these animals catch food.

The sea gooseberries solve this dilemma in a most peculiar way. When food is caught on the tentacles, it triggers the animal to begin rapidly somersaulting. This rolling action brings the tentacles across the mouth, where the lips grab the food and ingest it.

**Scientific name** *Pleurobrachia pileus*

**Phylogeny** PHYLUM Ctenophora / CLASS Tentaculata / ORDER Cydippida

**Notable anatomy** spherical body with two tentacles

**Position in water column** epipelagic, in open ocean and neritic zone

**Size** body to about 1.5 cm (0.6 in.) in diameter

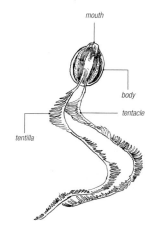

mouth

body

tentacle

tentilla

Distribution

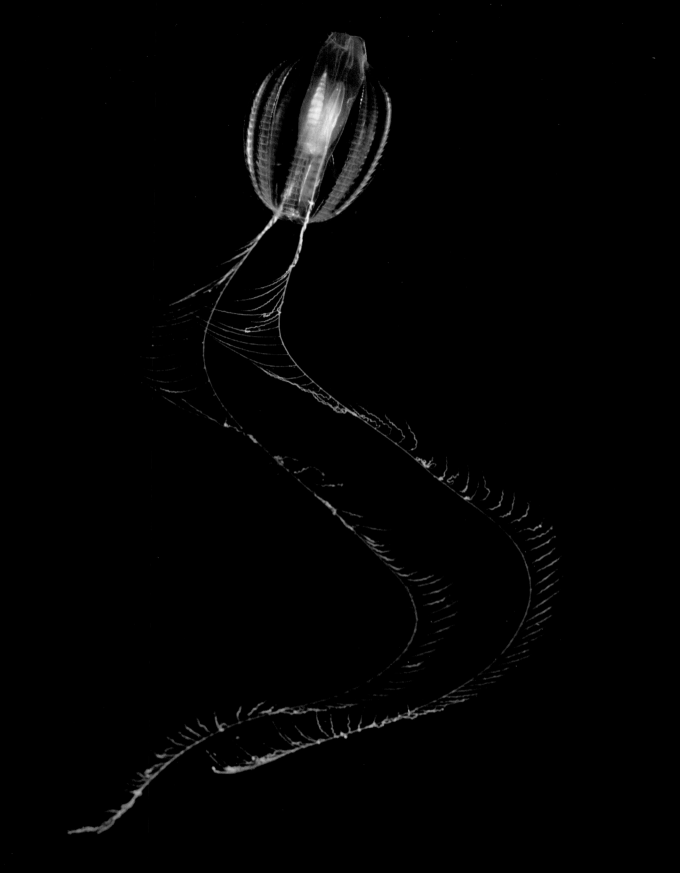

# LIZARD-TAIL JELLYFISH

Aτ first glance, the Lizard-tail Jellyfish (*Colobonema sericeum*) seems a humble species, lovely in its simplicity. It is about the size and shape of a small shot glass, and its body surface is sculpted into numerous longitudinal rows (such as those on a pumpkin). Its jelly is fairly firm and springy, even holding its shape out of water. Its color is translucent milky-white throughout; however, in the right lighting the smooth muscles in its body wall refract rainbows of color. The animal is not actually making the color; it is just an optical illusion resulting from the properties of the jelly in this species. Its mouth is a small, simple, short tube in the arch of the underside, and its stomach is an imperceptibly small cavity at its base. *Colobonema* has a few dozen long, fine tentacles arranged around the margin, which splay out when it is looking for a meal; it appears as a delicate swirl to some, or a fan-shaped death trap to others.

## The Lizard-tail Defense

The deep sea is a ferocious dog-eat-dog sort of world. Soft-bodied creatures such as jellyfish must rely on defensive ploys and escape strategies to avoid becoming someone else's dinner. Many species sting, some are transparent, and others flash bioluminescent light to startle their prey. *Colobonema* goes one step further. When necessary, it jettisons its tentacles, which once free of the body quickly begin glowing blue and wriggling like worms, providing a distraction so that the jelly can swim away undetected. It then regrows its tentacles and goes back to living its humble life. *Colobonema* is one of only a few jellyfish species known to use this defensive strategy, similar to that of lizards that escape the clutches of predators by ditching their tail tips.

**Scientific name** *Colobonema sericeum*

**Phylogeny** PHYLUM Cnidaria / CLASS Hydrozoa / ORDER Trachymedusae

**Notable anatomy** thimble-shaped, with many fine tentacles around the bell margin

**Position in water column** mesopelagic, in open ocean

**Size** bell to about 3 cm (1.2 in.) in height

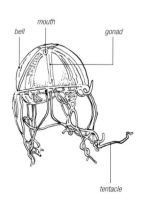

mouth

bell

gonad

tentacle

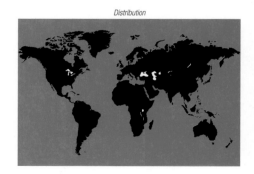

*Distribution*

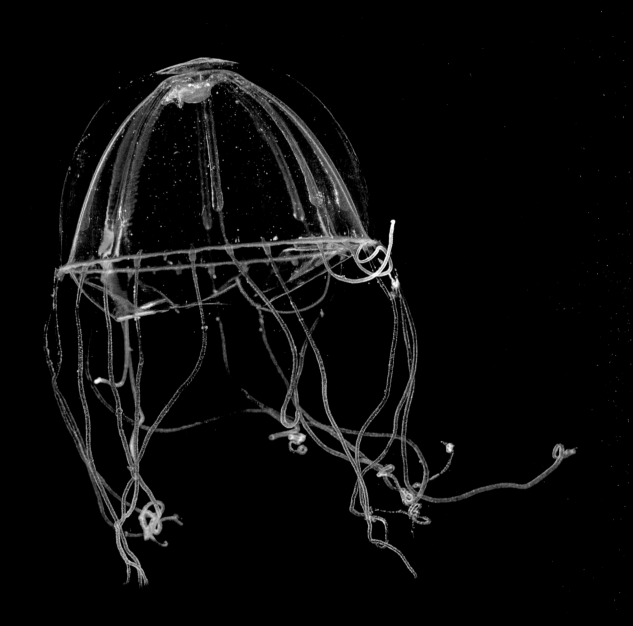

# UPSIDE-DOWN JELLYFISH

POSSIBLY THE MOST PECULIAR of all jellyfish are the upside-down jellies (genus *Cassiopea*), so named because of their lifestyle. *Cassiopea* is perfectly able to swim but instead spends most of its time resting on its back over the sediments. There, it whiles away the day, pulsating its bell every now and then to keep the water moving over its tissues. *Cassiopea* is a farmer rather than a hunter, and its tissues are packed with symbiotic algae that provide up to 90 percent of its nutritional needs. These algae are single-celled dinoflagellates called zooxanthellae, and are similar to the algae found in symbiotic relationships with corals.

*Cassiopea* has several unique adaptations to assist in its demersal—or bottom-resting—habit. Its bell is flattened, a clear advantage for stability on flat surfaces. Its arms are splayed out parallel to the body so they lay flat, for maximum sun exposure, rather than projecting perpendicularly away from the body, as is more typical for jellyfish that drift in the water and live with their mouths pointing downward. The surface of the oral arms is foliaceous, or leafy, providing ample surface area for the algae. Also, intriguingly, medusae may possess numerous tags attached to the arms; these are believed to be powerhouses of zooxanthellae for extra energy generation.

## Life at the Bottom

*Cassiopea*'s preferred habitat is shallow tropical lagoons and tidal pools, where it may be found in tremendous numbers, packed in cheek-by-jowl. In these locations, the abundance of jellyfish may be overlooked at first glance because of their resemblance to algae, but upon closer inspection, the bottom appears to be alive, quivering and pulsating.

**Scientific name** *Cassiopea* spp.

**Phylogeny** PHYLUM Cnidaria / CLASS Scyphozoa / ORDER Rhizostomeae

**Notable anatomy** flat bell with eight foliaceous arms

**Position in water column** resting upside down on sediments in very shallow coastal waters

**Size** bell up to about 30 cm (12 in.) in diameter

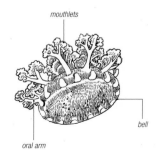

mouthlets

bell

oral arm

Distribution

# DEADLY BOX JELLYFISH

CHIRONEX FLECKERI, the Deadly Box Jellyfish, has the dubious distinction of being the world's most venomous animal. Its sting feels like a splash of boiling oil, searingly hot and indescribably painful. *Chironex* is the only animal that is capable of locking the heart in a contracted state, which it does in as little as two minutes. Once the heart is locked, resuscitation maybe ineffective and so early CPR is essential.

## Killer Tentacles

The box-shaped body, which grows to about 30 centimeters (a foot) in diameter, has up to 60 tentacles distributed in four groups around the perimeter—15 per corner. The tentacles are thick and flat, resembling tapeworms, and may extend to two or three meters (six to nine feet) long. *Chironex* trolls its tentacles through shallow tropical waters along sandy beaches looking for food. Young individuals prefer prawns,

but as the *Chironex* medusa grows, it undergoes a change in its food preferences. Older individuals actively hunt and subdue high-energy prey such as fish. The change is not merely behavioral: the stinging cells rearrange their dominance, and their toxin becomes more potent as well.

## Seeing the Light

There is another thing about the Deadly Box Jellyfish that makes it utterly fascinating—its eyes. Experiments have shown that *Chironex* will navigate toward a match lit across a dark room. Moreover, it is powerfully attracted to lights of all colors, swimming rapidly toward them, even from far away. However, it responds to blue light differently: it slows down and swims in a repeated figure-eight pattern through the light halo, with its tentacles streaming behind it fishing for food.

---

**Scientific name** *Chironex fleckeri*

**Phylogeny** PHYLUM Cnidaria / CLASS Cubozoa / ORDER Chirodropida

**Notable anatomy** box-shaped body with four groups of up to 15 tentacles each

**Position in water column** shallow subtidal waters over sandy beaches; estuaries

**Size** bell to 30 cm (12 in.) in diameter

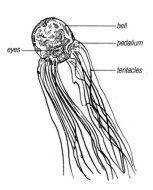

bell

pedalium

eyes

tentacles

*Distribution*

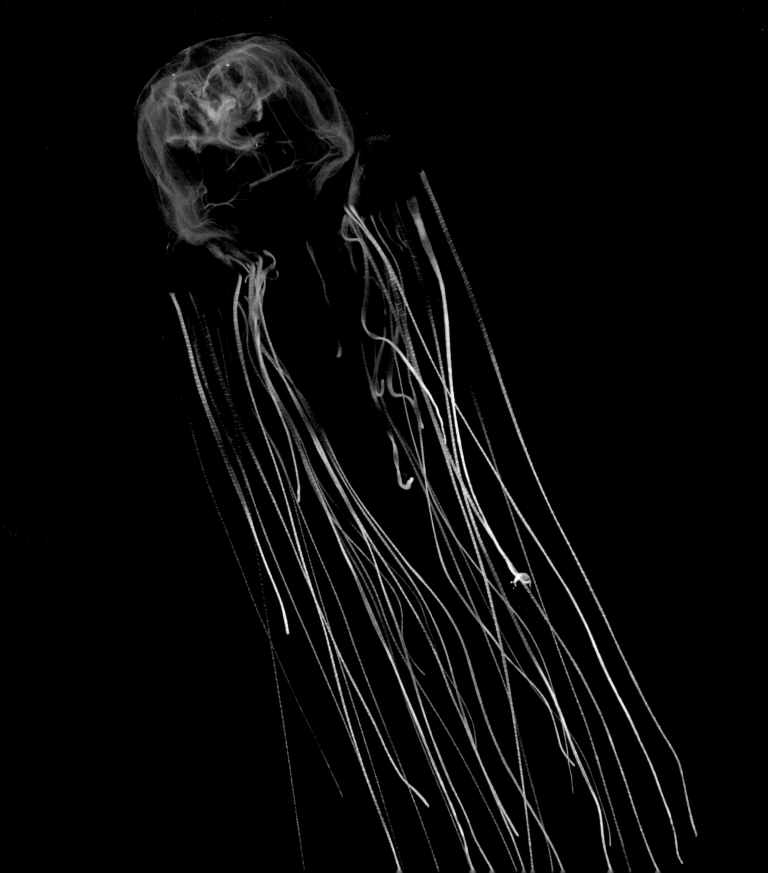

# LION'S MANE JELLIES

Normally, the lion's manes (genus *Cyanea*) look like a dinner plate with a mop underneath, but when the tentacles are relaxed to capture planktonic prey, the animal becomes a big, wild mass of tentacles, and sometimes it is difficult to perceive the body. All these structures create drag when the jellyfish is swimming. *Cyanea* solves this problem by having its tentacles attached entirely underneath the bell rather than at the margin, as in most other jellyfish. Even so, it still needs powerful muscles to give it the jet propulsion necessary to fight the current. These muscles are useful as a strong diagnostic for distinguishing species.

### How Many Lion's Manes?

It now appears that the lion's manes, once thought to be one widespread global species, *Cyanea capillata*, may be split into dozens of regionally distinct species. For example, around Tasmania, Australia, three new species of lion's manes were recently discovered, including one that can reach about one and a half meters (five feet) in bell diameter. However, in the North Atlantic, *Cyanea* can be nearly three meters (ten feet) in diameter, and their many hundreds of fine tentacles are said to reach 30 meters (nearly 100 feet) long.

### The Lion's Sting

Despite their unkempt appearance and sometimes massive size, most lion's mane species are fairly innocuous—more bark than bite—except for one. The red population found off the British Islands, currently classified as *C. capillata*, was the killer in a Sherlock Holmes murder mystery, "The Adventure of the Lion's Mane." There is nothing fictional about its sting, which causes a dangerous systemic effect. This suite of symptoms from jellyfish stings is known as Irukandji syndrome, which takes its name from a life-threatening species of box jellyfish, *Carukia barnesi* (page 154).

**Scientific name** *Cyanea* spp.

**Phylogeny** PHYLUM Cnidaria / CLASS Scyphozoa / ORDER Semaeostomeae

**Notable anatomy** broad, flat disk with hundreds of tentacles attached under the bell in horseshoe-shaped groups

**Position in water column** epipelagic, in neritic zone

**Size** bell up to 3 m (10 ft.) in diameter; tentacles to 30 m (98 ft.) long

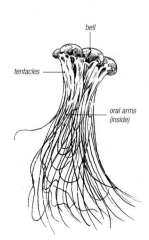

bell

tentacles

oral arms
(inside)

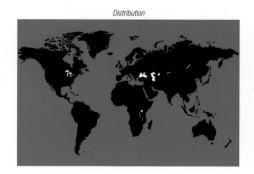

*Distribution*

CHAPTER TWO

# JELLYFISH
# LIFE HISTORY

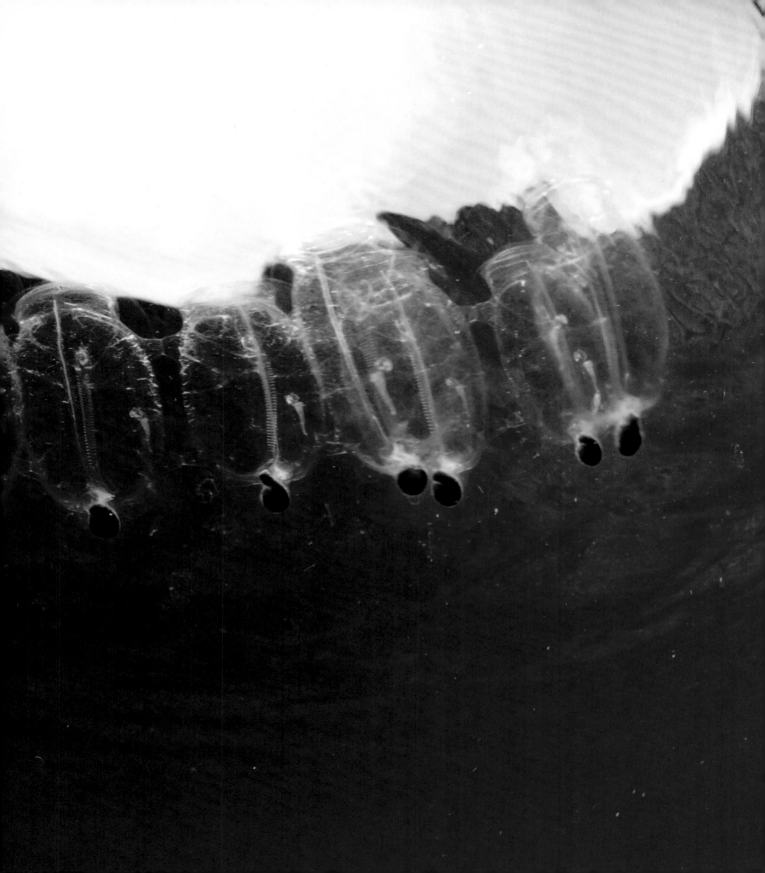

# INTRODUCTION TO JELLYFISH LIFE HISTORY

The strange world of jellyfish life history is both foreign and fascinating: it is utterly unlike anything we encounter in species that are more familiar to us. Cloning and hermaphroditism, both extraordinary in our everyday world, are not only prevalent in the jellyfish groups, they are so well developed there are many variations of each. The diversity of lifestyles, life stages, and body parts, can seem challenging to make sense of; however, they are the keys to how these simple creatures have survived for eons.

FOUR VERY DIFFERENT free-swimming jellyfish groups in three phyla are currently recognized (the medusae and siphonophores in phylum Cnidaria, the ctenophores in Ctenophora, and the salps and kin, or pelagic tunicates, in Chordata), plus two bottom-living groups (the stauromedusae in Cnidaria and the platyctenes in Ctenophora). Within these six different functional groups, however, the diversity of larval and adult stages, sexual and asexual stages, and benthic and pelagic stages is astonishing. All told, there are a couple of thousand species distributed among these jellyfish groups, and many of them have unique life history strategies.

## Jellyfish Life Stages

Most types of medusae have a minute, free-swimming primary larval stage, called a planula larva, though it differs dramatically in form, function, behavior, and duration across the major groups. Some species also have a secondary larval stage. The majority of jellyfish species during their life cycle alternate between a sexual stage and an asexual stage, and these often do not look anything alike. Historically, the alternating life stages of some species were even classified in entirely different higher taxonomic groups. For example, until recently the hydroids (asexual forms of hydrozoan medusae, of the phylum Cnidaria) were classified in two orders, whereas their corresponding hydromedusae (their sexual forms) were classified in five

orders; although these two classification systems overlapped in the biological sense, there was no overlap in the nomenclatural sense. For many years, this classification of the life history stages of the same species into two different taxonomic orders was standard practice. That classification system has since been fixed, but even today the majority of corresponding life history stages still have not been linked at the species level.

Moreover, across the jellyfish groups the same life history stages—such as sexual and asexual—have completely unrelated names, and corresponding morphological features such as mouths, muscles, canal systems, stomachs, and sensory organs, may also have different sets of terminology. These differences in jargon have come about through nonoverlapping sets of experts working in different taxonomic groups over the last several hundred years.

## Pelagic and Benthic Stages

Another aspect of jellyfish life history concerns their change of habitat during their different stages. In many species, the life cycle is split into pelagic (drifting) and benthic (seafloor) stages. These stages encounter different ecological conditions, seasonal changes, and predator/prey issues in the zones they inhabit. Thus, creatures that may be clone mates of one another have to navigate very dissimilar challenges and must deal with these pressures in divergent

ways. The genetics that underlie the species' varying structural adaptations and behavioral responses are a fascinating field of study, too.

Benthic stages produce adhesive substances for sticking to surfaces to stay on the sea bottom, while pelagic stages have bulked-up bodies or special structures to assist in flotation to stay off the sea bottom. Benthic stages often scrunch to avoid predators or encyst (enclose tiny bits of polyp tissue in a keratinized protective coating, allowing them to essentially hibernate) to escape undesirable conditions. Pelagic medusae, in contrast, use their muscles to orient themselves in the water column or to move directionally away from certain stimuli or toward others. Reproductive efficiency is enhanced by the aggregation of individuals into a cluster, and the different life history stages solve the problem of gathering together in quite dissimilar ways: benthic stages aggregate either by local asexual proliferation (cloning) or by some

means of attracting new planula larvae to settle, while pelagic stages aggregate passively with the aid of currents and windrows or actively by their attraction to certain conditions.

This chapter focuses on the strange life histories of the jellyfish groups in order of increasing complexity. Special attention is given to cloning and immortality, although without a doubt we have only scratched the surface of understanding these two phenomena. The permanently benthic (bottom-dwelling) forms of medusae and ctenophores are treated more thoroughly in "Anatomy of Benthic Forms" in chapter 1 (pages 18–19).

**Above left and right** *Phacellophora camtschatica*, commonly known as the Egg-yolk Jelly, is a large species found off the Pacific coast of North America. Like some other jellyfish, it uses a process called strobilation to link between the benthic polyp stage (above left), and the pelagic medusa stage (above right). The polyps act as a seed bank, developing long stacks of baby medusae to form huge blooms.

# CLONING 13 DIFFERENT WAYS

Generally, when we think of cloning, we conjure something odd and mysterious that happens in a laboratory, perhaps the creation of Franken-creatures in the test tubes of mad scientists. But almost all jellyfish are capable of cloning to an extent, some fabulously so. In fact, jellyfish can clone in at least 13 different ways.

## Polyp Replication

The most common form of jellyfish cloning is polyp replication. Hydroid colonies typically send out stolons, or runners, which sprout up new polyps at regular intervals. Scyphozoans also often bud from stolons, but these stolons are just a small finger of tissue that juts out and sprouts a new polyp, which then separates. More commonly, scyphozoans create new polyps by side-budding, which begins with a bulge developing on the column of the polyp, usually down near the base. This bulge becomes more pronounced, eventually sprouts tentacles and differentiates into a polyp; it then detaches and creeps away.

Another kind of budding that is commonly observed occurs by fragmentation. As polyps move around slowly, they leave tiny bits of tissue behind, which grow into new polyps. In a highly developed form of fragmentation, these tiny bits of tissue encyst, becoming covered by protective, keratinized, low rounded disks called podocysts. As the polyp moves forward, it leaves behind a trail of podocysts; these may hatch quickly, or they can persist for many years until conditions are favorable—at which time a polyp will emerge.

Cubozoan polyps not only clone replicates of themselves, they also bud off creeping polyps, which are long, wormlike polyps that glide around on cilia.

## TYPES OF CLONING

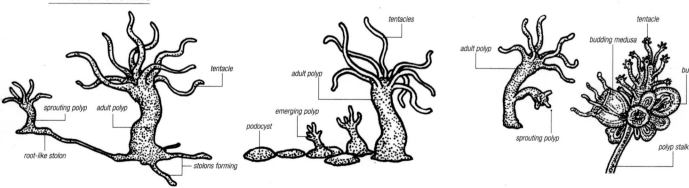

**Stolon budding** Many types of polyps are able to send out fleshy finger-like shoots or runners, called stolons. New polyps genetically identical to the parent polyp sprout up from these stolons. These are often used as a means of expanding the edge of the colony.

**Podocyst formation** Podocysts are tiny fragments of polyp tissue covered over by a keratinized dome. Some species, like the sea nettles, use these as a normal means of polyp replication, while other species, like moon jellies, use them to survive adversity.

**Polyp budding** This form of cloning is incredibly common. Simply, a clone sprouts from the side of the main stalk. Some species use this strategy to generate new daughter polyps, while other species use it as the primary means of budding off baby medusae.

## Medusa Production

The clonal production of the various classes of medusae by polyps is as different as the classes themselves. In scyphozoans, it occurs through strobilation, which is a process of segmentation and metamorphosis (described in "Scyphozoan Life History," pages 62–63). In some cubozoan species the process of medusa production transforms the entire polyp, so it is more metamorphic than clonal, because nothing is left over; in other species, the polyp remains after medusa production, which makes it a clonal process. In hydrozoans, medusa production typically occurs by budding, which occurs in a specific budding zone. Depending on the species, the zone may be near the base of the stalk or just below the mouth.

In some hydrozoan species, new medusae are budded from a parent medusa. This typically occurs on the gonad or in reproductive tissue around the manubrium, which essentially connects the mouth and stomach, but it may also occur along canals on the underside of the bell or along the margin of the bell at the base of the tentacles.

In *Turritopsis dohrnii*, the Immortal Jellyfish (page 74), new hydroids and medusae are produced through normal hydrozoan processes, but they sometimes follow from transformation of the medusa into a polyp, which is a unique type of cloning.

Some can clone by direct fission. The medusa spontaneously reshapes itself by pinching the bell margin inward at the middle, forming two sides that begin to separate. As the process continues, the mouth divides in two, each part splitting off with one of the half-jellies. In a variation of fission called schizogony, the medusa forms multiple stomachs and then splits into two daughter medusae.

While fission is a process wherein one jellyfish divides into two spontaneously, most jellyfish are also masters at repair and regeneration. For most species, being chopped in half, or even fourths, presents no problem: they just regrow the missing body parts in a matter of days.

## Other Cloning Methods

Additional methods of cloning are common, particularly in the siphonophores and the salps and kin, which are covered later in this chapter (pages 68–69 and 70–71, respectively).

Another type of reproduction, which blurs the distinction between sexual and asexual, is selfing. At least one ctenophore species, *Mnemiopsis leidyi* (page 196), is able to fertilize its own eggs with its own sperm; while this is technically sexual reproduction, because there is only one parent it is also technically cloning. Other siphonophores, however, don't seem to clone in the normal sense.

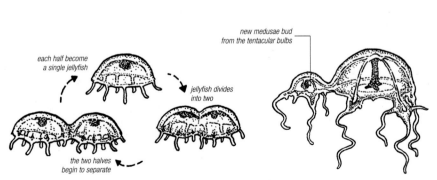

**Fission** Some medusae can use fission to clone. First the sides of the bell and the stomach spontaneously begin to pinch inward, forming an hourglass shape. The stomach splits into two first, and then the body shortly follows. The two halves each then grow into full medusae.

**Medusa budding** Some species of medusae can clone other medusae by budding from various bodily structures. In some, like *Niobia*, new medusae can sprout from the tentacle bulbs. Other species, like the minute, invasive *Eucheilota*, can bud new medusae from the gonads in

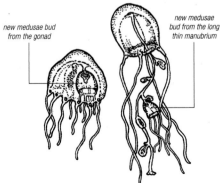

the arch of the bell; these developing buds may be too numerous to count. Still others, like *Stauridiosarsia*, can bud new medusae from their long, slender manubrium (throat or stomach). Sometimes these daughter medusae can also produce new buds.

# CTENOPHORE LIFE HISTORY

The simplest life history among the sea jellies is that of the ctenophores. More than 100 species have been described in the phylum Ctenophora, and nearly all aspects of the biology and ecology of most of them remain unknown. Particularly with the more delicate species, which are so diaphanous that even a gentle ripple in the water tears them into shreds, laboratory study lasting beyond a day or so has proven elusive. For those few whose life history has been studied, it is fairly simple. Individuals typically spawn their eggs and sperm; the fertilized eggs develop into a larval stage, which transforms into a small version of the adult form that then grows quickly and matures.

A REMARKABLE EVOLUTIONARY CHANGE between the cnidarian medusae and the ctenophores is seen in their symmetry. Whereas cnidarians are radially symmetrical—that is, they can be cut, as a pie, into numerous equal slices—ctenophores are strictly bilaterally symmetrical, which means there is only one way to slice them into two equal parts. Their bilateral symmetry is established in their embryonic development and continues throughout their lifetime.

All sorts of bizarre body forms are found in adult ctenophores: spheres with two tentacles; compressed globs with two large lobes; long ribbons; pockets, open along one edge; and flat creeping films. All start life as a tiny, spherical, free-swimming organism with eight short rows of cilia paddles and usually two tentacles; it is called a cydippid larva because of its resemblance to the adults in the order Cydippida, the sea gooseberries (page 44).

## Reproduction

Most ctenophores are hermaphrodites, which means an individual is both male and female. In the invertebrate world, hermaphroditism is often a normal part of an organism's life history. Ctenophores display several forms: the sexes may occur in the individual at the same time (simultaneous hermaphroditism), or individuals may be sequentially hermaphroditic, female first, then male (protogynous hermaphroditism), or male first, then female (protandrous hermaphroditism).

The gonads develop along the sides of a ctenophore's comb rows, the ovaries on one side and the testes on the other. Some ctenophores undergo two periods of sexual maturity, one in the larval form and the other in the adult, with the gonads degenerating in between. This unusual process is called dissogeny. All ctenophores except the benthic platyctenes shed their gametes (eggs and sperm) into the seawater, where fertilization takes place; platyctenes brood their young on the underside of the body.

In the order Cydippida, during the transformation to the adult form, the larvae undergo very few changes other than simply becoming larger and developing gonads (testes and ovaries) for sexual reproduction. In the orders Lobata (the lobates, e.g., *Bolinopsis*, *Mnemiopsis*, page 196, and *Ocyropsis*, page 40) and Cestida (the Venus's girdles, e.g., *Cestum* and *Velamen*, page 132), the cydippid larvae lose their tentacles and stretch into elaborate shapes. As the lobates grow, two large lobes develop on the oral (mouth) end; the Venus's girdles compress and elongate in the extreme, taking on a

## CTENOPHORE LIFE CYCLE

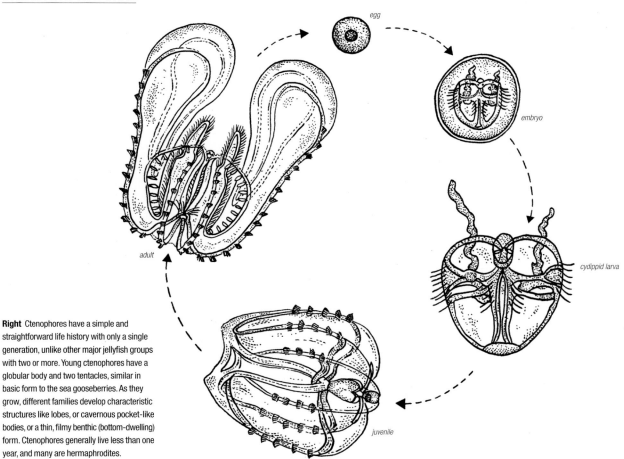

*egg*

*embryo*

*cydippid larva*

*adult*

*juvenile*

**Right** Ctenophores have a simple and straightforward life history with only a single generation, unlike other major jellyfish groups with two or more. Young ctenophores have a globular body and two tentacles, similar in basic form to the sea gooseberries. As they grow, different families develop characteristic structures like lobes, or cavernous pocket-like bodies, or a thin, filmy benthic (bottom-dwelling) form. Ctenophores generally live less than one year, and many are hermaphrodites.

ribbonlike form that can be more than a yard long. In the cydippid larvae of the class Nuda, which is defined by its lack of any trace of tentacles in all life stages, the primary change is an enormous expansion of the stomach as the animal grows.

### Reproduction in *Mnemiopsis leidyi*

The life history of one ctenophore species—*Mnemiopsis leidyi*, infamous for its fisheries-damaging potential as an invasive species outside its native western Atlantic range—is nothing less than astonishing. It is a simultaneous hermaphrodite that is able to self-fertilize, which means its own sperm fertilizes its own eggs. Usually, natural barriers

are built into a species' reproductive cycle so self-fertilization is not possible. These barriers include differences in timing (day versus night spawning) or in buoyancy (floating versus sinking) of the sex cells. *Mnemiopsis* begins laying eggs within 13 days of its own birth, and can lay up to 10,000 eggs a day throughout its several-month lifetime.

In addition to breeding and cloning, ctenophores have remarkable regeneration capacity. Experiments with *Mnemiopsis* have shown that it can be cut in halves, thirds, and even quarters, and still replace missing body parts in just a few days and resume life as normal.

# SCYPHOZOAN LIFE HISTORY

Members of the cnidarian class Scyphozoa, often referred to as the "true jellyfish," have perhaps the most straightforward life histories of any of the medusa groups. Even so, their developmental cycles are remarkable, and unfamiliar to most of us.

## Scyphozoan Life Stages

The scyphozoan medusae that we see are generally either male or female; however, each is part of a clone batch that may contain both sexes. The medusa is the sexually reproducing stage and is a unit of dispersal. It is comparable to the individual fruit on a tree, such as an apple or a pear. When a fruit breaks off the branch and rolls away from the tree, it takes its seeds to a new location. Jellyfish medusae do the same thing.

What, one may wonder, is the counterpart of the tree—the part that produces the fruit? Jellyfish polyps are small and inconspicuous, but in many ways they are the most important part of the life cycle. Just as without trees there would be no fruits, without polyps there would be no medusae.

Many people think of polyps as larvae and medusae as adults, but this is not quite accurate. The only true larval stage in the life cycle—that is, a larva resulting from the union of sperm and egg—is the planula larva. The scyphozoan planula is a tiny, ciliated cigar-shaped creature produced by fertilization between male and female medusae. These larvae swim in corkscrew patterns for a few days until they find a suitable rock or branch to settle on; then they metamorphose into a polyp.

## Scyphozoan Polyps

Properly called scyphistomae (singular, *scyphistoma*), scyphozoan polyps are typically wineglass-shaped or fluted, and range from less than a millimeter (0.04 inch) to nearly a centimeter (less than half an inch) tall, depending on species. They stick to rocks, shells, or leaves by means of a sticky foot at the bottom of the body. Around the top of the body is a crown of tentacles, with a mouth in the middle. These polyps are voracious consumers of just about every little particle of plankton or organic matter that passes their way. This huge appetite sustains the polyps' more or less constant process of cloning replicates of themselves.

Polyps can form by way of two different growth and development routes: primary polyps form from metamorphosis of the planula larva; secondary polyps form by budding from an existing parent polyp. Budding may come about via a range of pathways: most scyphozoan polyps reproduce by side budding; however, some species, such as the sea nettles (*Chrysaora* species), develop podocysts, encased fragments of tissue that later "hatch," as their primary form of clonal proliferation (these methods are explained in "Cloning 13 Different Ways," pages 58–59).

Scyphozoan polyps live all year-round, and in fact can live for many years. The oldest known polyp colony (based in a laboratory) dates back to 1935 and has been reproducing clonally since that time. When conditions are right, usually once each spring and fall, the polyps undergo a fascinating process to bud off baby jellyfish. During this process, called strobilation, the polyp elongates and differentiates into a stack of disks, resembling a stack of tiny saucers. At first, the polyp—called a strobila during this process—still looks like a polyp, though with a series of several to dozens of ribs where the stalklike body is normally found between the polyp's crown and base. Each rib develops into a daisy-like ephyra (plural, *ephyrae*); these are the developing jellyfish. During strobilation,

## SCYPHOZOAN LIFE CYCLE

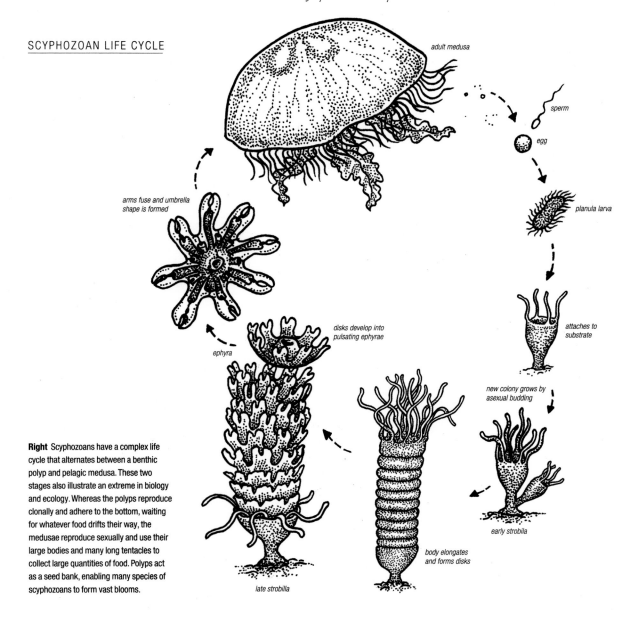

*adult medusa*

*sperm*

*egg*

*arms fuse and umbrella shape is formed*

*planula larva*

*attaches to substrate*

*new colony grows by asexual budding*

*disks develop into pulsating ephyrae*

*ephyra*

*early strobila*

**Right** Scyphozoans have a complex life cycle that alternates between a benthic polyp and pelagic medusa. These two stages also illustrate an extreme in biology and ecology. Whereas the polyps reproduce clonally and adhere to the bottom, waiting for whatever food drifts their way, the medusae reproduce sexually and use their large bodies and many long tentacles to collect large quantities of food. Polyps act as a seed bank, enabling many species of scyphozoans to form vast blooms.

*body elongates and forms disks*

*late strobilla*

the polyps resorb their tentacles and are unable to feed; after strobilation, they resume gorging themselves with plankton, readying for future strobilation and budding events.

Ephyrae typically have eight petallike arms, each with a tiny, shiny sense organ at the end. As the ephyrae develop on the strobila, they begin to pulsate and, after a few days, one by one they break free and swim away. They feed voraciously on plankton; after only a few days the arms fuse and form an umbrella shape, and the remaining structures develop into what we recognize as the components of a medusa.

The juvenile medusa eats constantly and grows rapidly. Many species are strobilated in the spring as the weather warms, then they grow, reproduce, and die by fall. In that time, some may reach incredible body sizes, up to the size of trash can lids.

# CUBOZOAN LIFE HISTORY

The life history of cubozoans, or box jellyfish, members of the cnidarian class Cubozoa, is similar to that of many other jellyfish classes, in that it has a fixed polyp stage and a free-swimming medusa stage; however, many aspects are unique to the class.

## Cubozoan Life Stages

In cubozoans, the sexes are always separate in the medusa stage; hermaphroditism is unknown in the group. Most species spawn their eggs and sperm out into the sea, but in a few species the female retains the eggs and fertilization occurs internally. In *Tripedalia cystophora* (page 122) and *Copula sivickisi* (page 88) the females brood their embryos after fertilization.

The cubozoan planula larva is teardrop-shaped and bears a number of small dark eyespots around the broadest part of the large end. The planula swims in a spiral pattern by means of cilia, with its large end forward. After spending a couple of days swimming, it settles on the underside of a hard object such as a shell or rock and transforms into a cubopolyp. Cubopolyps are immediately distinguishable from all other types of polyps by the presence of one or several nematocysts embedded in the tip of each tentacle.

## Polyp Development

The primary polyp (the cubopolyp formed from the planula larva) produces creeping polyps. These are very slender and elongate—up to about a centimeter (almost half an inch) long—with a mouth at one end encircled by a crown of several tentacles. They glide around mouth-first for a few days before they finally attach and transform into a secondary polyp.

The secondary polyps grow to one millimeter (0.04 inch) tall, and are fixed to hard surfaces by means of a sticky foot. They bud throughout their life. Under certain seasonal conditions, which are not yet explicitly known for most species, the secondary polyps undergo a radical transformation during which the whole polyp develops into a baby cubomedusa. This is very different from the metamorphic process in the Scyphozoa, in which the polyp undergoes a process of strobilation, forming a stack of tiny disks (ephyrae) that bud off, leaving the original polyp behind. Rarely, the cubopolyp remains as a remnant, but this appears to be the exception rather than the rule. Moreover, only one baby medusa is produced by a cubopolyp, compared to the many ephyrae, which in turn become medusae, produced by most scyphopolyps. This numerical difference probably has a strongly limiting effect in the bloom capacity of cubozoans.

## Metamorphosis

The transformation process of cubozoans is remarkably complex, but as yet is well understood for only a couple of species. When the polyp begins to transform, the simple rounded shape encircling the mouth takes on a more quadrangular (four-cornered) shape. Then the tentacles, which are normally equally distributed around the rim of the polyp, congregate at the four corners. The tentacles resorb, and their bases coalesce and thicken into knobs, which develop small pigment spots. These knobs become the rhopalia, or sensory structures, in the adult, and the pigment spots develop into eyes. The metamorphosing polyp grows four new tentacles between the four rhopaliar knobs. These four primordial tentacles become the pedalia—cartilaginous "legs" on which the tentacles or tentacle clusters are later borne. At tropical water temperatures, metamorphosis from polyp to medusa takes about four to five days; the onset of metamorphosis may occur about 10 to 12 weeks after the polyp forms.

## CUBOZOAN LIFE CYCLE

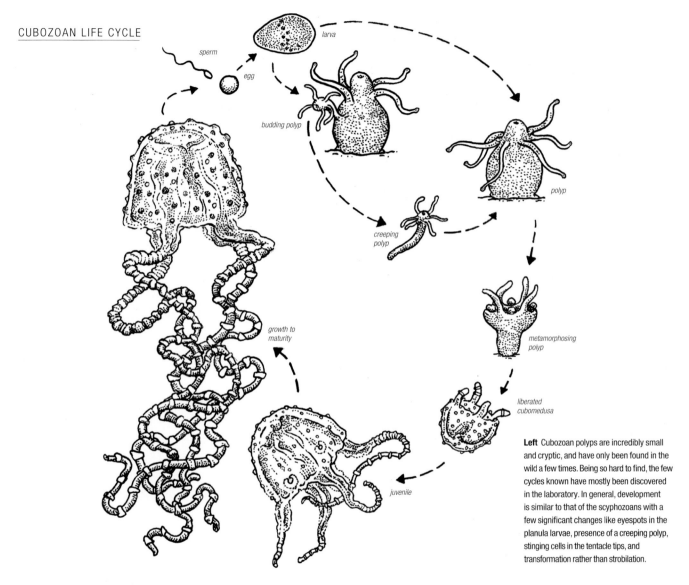

*sperm*

*egg*

*larva*

*budding polyp*

*creeping polyp*

*polyp*

*metamorphosing polyp*

*liberated cubomedusa*

*growth to maturity*

*juvenile*

**Left** Cubozoan polyps are incredibly small and cryptic, and have only been found in the wild a few times. Being so hard to find, the few cycles known have mostly been discovered in the laboratory. In general, development is similar to that of the scyphozoans with a few significant changes like eyespots in the planula larvae, presence of a creeping polyp, stinging cells in the tentacle tips, and transformation rather than strobilation.

The liberated cubomedusa is about one millimeter (0.04 inch) tall. Growth rates are variable among species. In the laboratory, *Tripedalia* takes about 10 to 12 weeks to reach full maturity, at which time it is about one centimeter (less than half an inch) in diameter. In nature, the Deadly Box Jellyfish, *Chironex fleckeri* (page 50), grows about one millimeter (0.04 inch) a day, reaching a full size of about 30 centimeters (12 inches) in diameter.

Each cubozoan rhopalia contains a single large balance stone, called a statolith, which the animal forms by adding daily growth rings. With careful laboratory preparation, these statoliths can be ground, in a manner similar to the polishing of gemstones, so the growth rings can be counted. Only a couple of species have so far been aged by this method; both have revealed that cubozoans may live about two years. Most specimens in nature, however, probably live less than one year.

# HYDROZOAN LIFE HISTORY

The cnidarian class Hydrozoa is an incredibly diverse group, encompassing both solitary and colonial forms. Some species have both hydroid and medusa stages, while others have either one or the other. There are shallow-water hydrozoans, others restricted to the deep sea, and a few confined entirely to freshwater. No single life history strategy defines this class.

THREE MAJOR GROUPS of hydrozoans are recognized. The most numerous and familiar are those considered the "traditional hydrozoans" in the subclass Hydroidolina, which contains the species that have a dominant hydroid and may also have a medusa. The second group, the subclass Trachylina, contains the species with a dominant medusa and generally without a polyp, as well as a few highly unusual species. The third major group encompasses the siphonophores, currently classified as order Siphonophora within the Hydroidolina, though sometimes treated as a separate subclass; the life history of these unusual creatures is covered on pages 68–69. In this section, references to the Hydroidolina do not include the Siphonophora.

In hydrozoans, the hydroid, or hydrozoan polyp, acts as an asexual proliferation stage, creating replicates of itself through cloning. The medusa acts as a sexual dispersal stage for the reproductive cells (sperm and eggs). In species in which the hydroid does not bud off a medusa, the hydroid itself acts in both asexual and sexual capacities. Some species lack a hydroid. Occasionally the medusa may reproduce both sexually and clonally (asexually).

## Subclass Hydroidolina

In the Hydroidolina, most commonly the male and female medusae spawn openly in the water; the resulting embryo develops into a ciliated, mouthless, elongate teardrop-shaped, free-swimming planula larva. The larva typically swims for a few hours to a few days, then attaches itself by its anterior end to a hard surface and grows into a hydroid. The posterior end of the larva becomes the mouth and tentacles. These primary polyps grow and bud into colonies. Young medusae are budded from the colonies, a process often triggered by changes from day to night or vice versa.

Within the Hydroidolina, there are two further main divisions based on the presence or absence of hard, cuplike protective structures on the colony called thecae (singular, *theca*). The thecate hydroids (order Leptothecata) possess such a structure, whereas the athecates (order Anthoathecata) lack it. Both thecate and athecate hydroids produce medusae that are generally built on a tetraradial (four-parted) body plan, but the two groups are quite different in other respects. Thecate medusae tend to be fairly flat and transparent; examples include *Aequorea victoria* (page 198) and *Obelia* species (page 204). Athecate medusae are typically more bell- or turret-shaped and are often brightly colored; examples include the Immortal Jellyfish (*Turritopsis dohrnii*, page 74) and *Polyorchis penicillatus* (page 172).

## Subclass Trachylina

The Trachylina is split into several divisions; the majority of species are found within the orders Trachymedusae and Narcomedusae. Both are typically found in the open ocean or deep sea and lack a conventional polyp. The trachymedusae are deeply bell-shaped and typically have an octoradial (eight-

## HYDROZOAN LIFE CYCLE

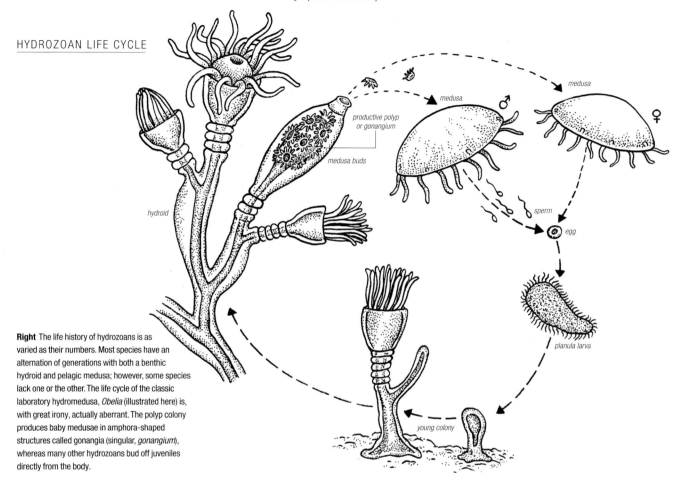

productive polyp
or gonangium

medusa buds

medusa

medusa

hydroid

sperm

egg

planula larva

young colony

**Right** The life history of hydrozoans is as varied as their numbers. Most species have an alternation of generations with both a benthic hydroid and pelagic medusa; however, some species lack one or the other. The life cycle of the classic laboratory hydromedusa, *Obelia* (illustrated here) is, with great irony, actually aberrant. The polyp colony produces baby medusae in amphora-shaped structures called gonangia (singular, *gonangium*), whereas many other hydrozoans bud off juveniles directly from the body.

parted) symmetry; examples include the Lizard-tail Jellyfish (*Colobonema sericeum*, page 46). The narcomedusae, such as *Aegina citrea* (page 84), often lack explicit symmetry and are distinguished by solid, rigid tentacles that are attached on the topside of the bell.

In the order Narcomedusae, the planula larva typically develops into a second larval stage, called an actinula larva, which then transforms into a medusa. The actinula looks like a short, stalkless polyp. In some species, the planula develops into a form known as a degenerative medusa, which then produces parasitic larvae. Different species of narcomedusae parasitize different host jellyfish species. These parasitic larvae then flatten into a stolon that in turn buds off medusae.

The complete life history is unknown for most species in the order Trachymedusae. However, those that are studied typically develop directly from the planula larva into a juvenile medusa without a polyp stage.

Freshwater species of the Trachylina—for example *Craspedacusta sowerbii* (page 162)—have tiny polyps that grow on underwater vegetation and hard structures such as rocks or tree branches on the bottoms of streams and ponds. Each medusa is either male or female; however, a body of water is often infested with only one sex. Upon emerging, *Craspedacusta* planula larvae swim for a few hours, then lose their cilia and settle to the bottom. After a few days the larvae develop into tiny polyps, each about one millimeter (0.04 inch) tall.

# SIPHONOPHORE LIFE HISTORY

The siphonophores, currently classified as order Siphonophora within the hydrozoan subclass Hydroidolina, are certainly among the most fascinating of all groups in the animal kingdom. A debate has raged for centuries over whether the siphonophore is truly an individual or truly a colony. Compelling arguments exist for both perspectives. Perhaps siphonophores are just too strange to be shoehorned into such neat categories. However, for the sake of simplicity, we shall refer to them here as colonies.

## Siphonophore Suborders

There are three suborders of siphonophores. Those of the suborder Calycophorae have swimming bells (nectophores) but lack a float; Giant Heart Jelly (*Praya dubia*, page 38) belongs to this group. These siphonophores have two parts to their life cycle; often their forms during these two parts look nothing alike, and some of these life stages have been classified as different species. In its asexual stage, the complete animal is generally referred to as the polygastric stage; this may be thought of as the "fully grown" siphonophore. The sexual stage is called a eudoxid (*you-DOCKS-id*); in some species this is a large and conspicuous creature, while in others it may be much smaller.

The other two siphonophore suborders are the Physonectae, which have a gas-filled float and swimming bells (examples are *Physophora*, page 156, and *Nanomia*); and the Cystonectae, siphonophores with a gas-filled float but no swimming bells (such as the Portuguese Man-of-war, page 34, and *Rhizophysa* species, page 206). In these two suborders the eudoxid stage is inconspicuous.

## Structure of the Colony

All siphonophores are free-floating or free-swimming, and each colony therefore appears to be a single organism. The components of the colony are so specialized in their form and function that they barely resemble one another. The members are differentiated into what are known as "persons"—swimming persons, defensive persons, digestive persons, reproductive persons, and so on. In all, seven different types of persons are present in each colony. These persons act for the benefit of the colony as a whole, and they cannot live or function on their own.

The different types of persons occur together in groups, each called a cormidium (plural, *cormidia*), along the main stem of the colony. These cormidial groups are added as the colony grows.

Each colony is sectioned into two structural and functional parts. The upper stem portion is called a nectosome; this is where the swimming bells (nectophores) are located, which may be thought of as rudimentary medusae. The nectosome supports and propels the colony. The lower stem portion is called a siphosome; this is where the colony members live: the feeding persons, reproductive persons, stinging persons, and so on. The float, in those species possessing one, is located at the very top of the colony, above the nectosome.

Some species bud extra colony members from a single budding zone at the top of the stem, while other species may bud from numerous budding zones along the stem. In many species the nectophores are replaceable and will quickly regrow if lost. In a few species, however, such as *Muggiaea* and *Sphaeronectes*, loss of the nectophore is terminal.

## SIPHONOPHORE LIFE CYCLE

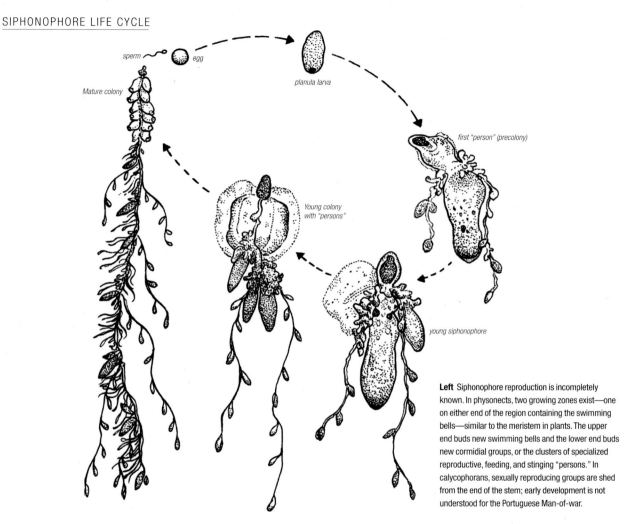

*sperm* *egg*

*planula larva*

*Mature colony*

*first "person" (precolony)*

*Young colony with "persons"*

*young siphonophore*

**Left** Siphonophore reproduction is incompletely known. In physonects, two growing zones exist—one on either end of the region containing the swimming bells—similar to the meristem in plants. The upper end buds new swimming bells and the lower end buds new cormidial groups, or the clusters of specialized reproductive, feeding, and stinging "persons." In calycophorans, sexually reproducing groups are shed from the end of the stem; early development is not understood for the Portuguese Man-of-war.

### Siphonophore Reproduction

While the morphological or structural variability of siphonophores has been long and widely appreciated, their life history strategies are likely just as variable. However, so little is known about the life history of most species that a generalized account is impossible.

A fully grown siphonophore is like a giant larva: it does not become a sexually mature individual itself but rather buds off eudoxids, the medusa-like sexual stage, which then produce larvae that grow into the recognizable siphonophore. Most species are monoecious, which means they bud off both male and female members, but some species are dioecious, and each colony buds off either one sex or the other. The most familiar siphonophore, the Portuguese Man-of-war, has separate sexes.

In siphonophores, female persons produce yolky eggs. Large quantities of eggs and sperm are broadcast out into the sea, where a few of each will meet and achieve fertilization. The resulting larva develops into the asexual colony.

The life span of individual siphonophores is poorly studied, but is believed to be up to 10 or more years, though most probably live for less than a year.

# LIFE HISTORY OF SALPS AND THEIR KIN

Pelagic tunicates, which include the salps, pyrosomes, and their kin, have the most complicated life histories of all the jellyfish. They are also the most evolutionarily advanced, belonging to the same phylum, Chordata, as humans. Pelagic tunicates possess a notochord at some point during their larval development; this is the evolutionary precursor to the vertebral column or backbone that characterizes vertebrates, including humans. Even though these creatures look like jellyfish as adults, their embryology clearly links them more closely to humans than to the other jellyfish groups.

## Salps

The most common pelagic tunicates are strange barrel-like creatures called salps, which belong to the order Salpida. Most salps alternate between two main life stages: an aggregate stage of numerous colony members, or zooids, chained together, and a solitary stage of a single zooid. All salps are fully pelagic throughout their lifetime; there is no polyp stage. The solitary stage reproduces asexually, developing a chain of aggregates from a budding stolon. The youngest members of the chain are those developing nearest the parent; the oldest are at the far end. The aggregates reproduce sexually and are sequentially hermaphroditic; the eggs of younger female chains are fertilized by sperm produced by older males from other chains. The embryo develops inside the atrium of the parent's barrel-shaped body, where it attaches to and obtains nutrition from the body wall. When development is complete, the young salp breaks free from the parent and becomes free-swimming, developing its own stolon and budding chain.

The growth rate of salps is simply phenomenal. At least one species, the Common Salp (*Thalia democratica*, page 208), can grow up to 10 percent of its body length *per hour*, and go through two generations in a day. Indeed, the name *Thalia* is Greek for "blooming."

## Pyrosomes

Pyrosomes, such as the Fire Bodies (*Pyrostremma spinosum*, page 42), are essentially colonial salps in which the zooids, or individual colony members, are embedded perpendicularly in a gelatinous matrix. The colony serves the functions of both the sexual stage and the asexual stage. Pyrosomes, which belong to the order Pyrosomatida, have a reproductive sequence similar to that of the salps. The pyrosome itself grows by asexual budding, whereas each zooid reproduces sexually by developing its own egg or sperm. The embryo developing in the atrium of the parent zooid develops four primordial buds; when the developed embryo becomes free of the parent, the original embryonic individual degenerates and the four buds then form a new pyrosome colony by secondary budding.

## Doliolids

Within the pelagic tunicates group, the doliolids (order Doliolida) have by far the most complicated life cycle, with up to six distinct body forms among multiple life stages. Doliolids are very small, barrel-shaped organisms that are distantly related to salps. They are easily distinguished from salps—even very young ones—by their completely encircling, parallel muscle bands, which lack the breaks in

SALP LIFE CYCLE

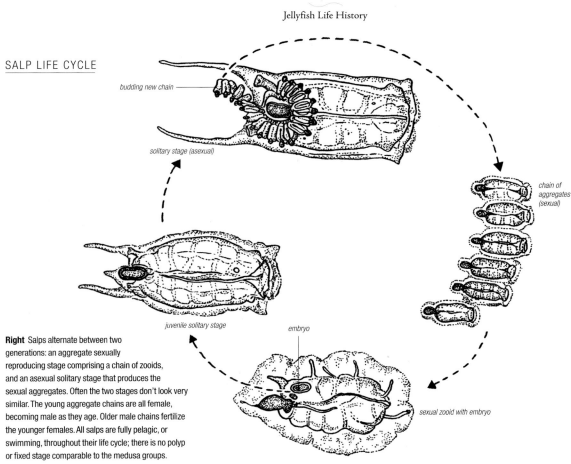

*budding new chain*

*solitary stage (asexual)*

*juvenile solitary stage*

*embryo*

*chain of aggregates (sexual)*

*sexual zooid with embryo*

**Right** Salps alternate between two generations: an aggregate sexually reproducing stage comprising a chain of zooids, and an asexual solitary stage that produces the sexual aggregates. Often the two stages don't look very similar. The young aggregate chains are all female, becoming male as they age. Older male chains fertilize the younger females. All salps are fully pelagic, or swimming, throughout their life cycle; there is no polyp or fixed stage comparable to the medusa groups.

the bands that salps have. After hatching, a young doliolid oozooid (solitary zooid) forms a threadlike tail, called a cadophore; trophozooids (nutrition zooids) sprout from both sides of the cadophore and, by feeding on phytoplankton, supply food to the whole colony. As the trophozooids grow, the oozooid, known as a nurse at this stage, grows too, and more trophozooids are budded. As the number of trophozooids surpasses five, the digestive tract and other organs of the nurse degenerate, and it becomes entirely reliant on the trophozooids for nutrition; as the number exceeds 20, small carrier zooids called phorozooids are sprouted. Phorozooids in turn sprout hundreds of gonozooids, the sexual stage, over their weeklong life span. The gonozooids are hermaphrodites: eggs are spawned approximately every second day, while sperm is spawned

intermittently. Doliolids capture phytoplankton with a mucous feeding filter, as salps do; in conditions of plentiful food a single doliolid can produce thousands of offspring within a few days, leading to blooms of more than 1,000 individuals per cubic meter (about 35 cubic feet) of water.

**Appendicularians**

Members of the class Appendicularia, such as *Oikopleura* spp. (page 82), have the least complicated life history in the pelagic tunicate group. Appendicularians (also called larvaceans) are neotenic, which means their larval characters are retained in the adult stage; they resemble the tadpole-shaped larvae of benthic tunicates. Typically, an embryo formed by reproduction between male and female parents develops directly into a young larvacean.

# THE SECRET TO IMMORTALITY

Mortal beings, we humans have a limited life span. We are born, we live, and then we die. Normally, when an organism dies, microbes begin decomposing the cells and liberating the molecules to be used by other organisms. The carbon, nitrogen, and other nutrients that make up our bodies have been used by many organisms before us, and will be used again by many others afterward. This is the cycle of life. Well, at least for most organisms.

ONE OF HUMANITY's ongoing quests has been the search for eternal life. It was a surprise, therefore, when scientists discovered that a jellyfish, the Mediterranean species *Turritopsis dohrnii* (page 74), has already found the means for bypassing death.

Almost all jellyfish species are immortal in one sense, because they are able to reproduce clonally in their polyp stage. Long after the bodily organism is gone, its complete genetic identity is still intact in another bodily organism, whether that is in another polyp in the colony or a medusa budded off by it. By analogy, imagine if a person were to chop off his or her hand, and the hand then grew a new body; even when the original person passed on, the new body that arose from the hand would still survive.

Moreover, many jellyfish are known for their remarkable regenerative ability; they can regrow missing portions from small parts of themselves when they have been cut into halves or even fourths or eighths. But *Turritopsis* goes one step further. When the bodily organism dies, it does not cease to exist: it comes back, transformed.

### The Transformation Process

Upon dying, *Turritopsis dohrnii* keeps its own nutrients and recycles them within its own little cycle of life. When the medusa is subjected to mechanical damage, a sudden temperature rise, or other stress, it goes through a characteristic transformation process. The bell and tentacles degenerate and seem to disappear, but the dissociated cells from both the bell and the circulatory radial canal system reaggregate within a few days. First stolons and then polyps appear. In other words, as the medusa dies and the tissues begin to break down, instead of decomposing, the cells reform as hydroids; this transformation process is known as transdifferentiation. The new polyps then proliferate into colonies and bud new medusae, just as jellyfish usually do. Experiments have shown that this occurs at an ambient summer temperature (22°C [72°F] or above); lowering the culture temperature causes the process to be suspended until it warms back up again.

The key stimulus in this process appears to be stress, as unstressed medusae are not known to undergo this transformation. Researchers have been able to raise *Turritopsis* through at least 10 iterations of this cycle in less than two years, so it appears that it may go on indefinitely. Moreover, all stages of growth can undergo this process, from newly released juvenile medusae to fully mature specimens. This raises all sorts of questions about the long-term bloom capacity of this species, particularly given the invasive capacity of jellyfish in stressed environments.

Close relatives are found in South Carolina, New Zealand, and southern Australia. Whether these other species are also immortal is not yet known.

## *TURRITOPSIS DOHRNII* LIFE CYCLE

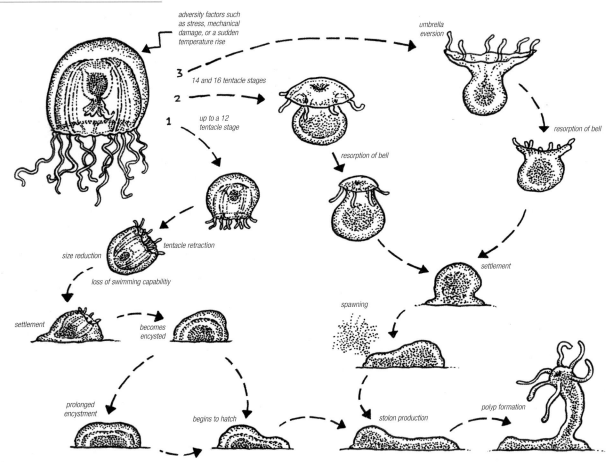

adversity factors such as stress, mechanical damage, or a sudden temperature rise

umbrella eversion

**3**

**2**  *14 and 16 tentacle stages*

**1**  *up to a 12 tentacle stage*

*resorption of bell*

*resorption of bell*

*tentacle retraction*

*size reduction*

*loss of swimming capabilitiy*

*settlement*

*settlement*

*becomes encysted*

*spawning*

*prolonged encystment*

*begins to hatch*

*stolon production*

*polyp formation*

### Aging in Reverse

Another fascinating aspect of *Turritopsis*'s immortality is in the genetics behind it. Human aging is unidirectional—that is, our cells are programmed to age in one direction, from young to old. Different processes are triggered at certain milestones in our cells' lifetime and our bodily lifetime. We go through puberty roughly a fifth of the way through our life. We stop growing at roughly the one-quarter mark. Women go through menopause after the halfway point. Our skin ages, but it does not un-age. *Turritopsis*'s genes go through comparable unidirectional processes, but then they reverse and essentially become young again.

**Above** *Turritopsis dohrnii* is the first known species to be truly biologically immortal. In response to physical stress, the medusa degenerates by any of three different pathways. The cells then reaggregate and regenerate as polyps. The cells reverse their genetic programming and become "young" again.

At first glance, this transformation in *Turritopsis* may seem similar to a caterpillar metamorphosing into a butterfly, but the resemblance is superficial. The caterpillar is a larval organism, and the butterfly is an adult, and the butterfly needs to mate to produce more caterpillars. *Turritopsis* is the developmental equivalent of a butterfly turning back into a caterpillar or an apple into a tree.

# IMMORTAL JELLYFISH

I have immortal longings in me.

WILLIAM SHAKESPEARE, *Antony and Cleopatra*

Tʜᴇ ǫᴜᴇsᴛ ꜰᴏʀ ᴛʜᴇ sᴇᴄʀᴇᴛ of immortality has driven mankind since time immemorial. It seems unlikely that in all this time—in spite of all the minds that have pondered it, the songs, the poems, and the prayers—anyone ever for a moment dreamed that this most splendid secret would be found in a jellyfish.

## How to Live Forever

The diminutive but fascinating *Turritopsis dohrnii*, the Immortal Jellyfish, grows only to the size of a pea; its fairly inconspicuous body is shaped like a thimble, with dozens of long, fine tentacles radiating from the margin. But this tiny creature holds the spectacular distinction of presenting the world's first example of true biological immortality (discussed in detail in "The Secret to Immortality," pages 72–73).

*Turritopsis* has—in common with many other species of jellies—a free-swimming sexual medusa stage and a benthic asexual polyp stage. When the medusa dies, instead of disintegrating and decaying away as most other living things do, its cells actually reaggregate and reform into a new polyp colony through a cellular developmental process called transdifferentiation. It is thus capable of reverting to an immature life stage after reaching sexual maturity. The polyps then begin the life cycle over again, and are capable of budding off new baby medusae. This is the equivalent of a frog dying, and having its cells reaggregate back into a tadpole!

**Scientific name** *Turritopsis dohrnii*

**Phylogeny** ᴘʜʏʟᴜᴍ Cnidaria / ᴄʟᴀss Hydrozoa / ᴏʀᴅᴇʀ Anthoathecata

**Notable anatomy** small thimble-shaped bell with many fine tentacles; bright red inside

**Position in water column** epipelagic, in neritic zone

**Size** bell less than 1 cm (0.4 in.) in height

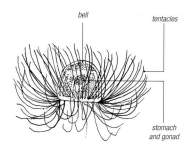

bell

tentacles

stomach and gonad

Distribution

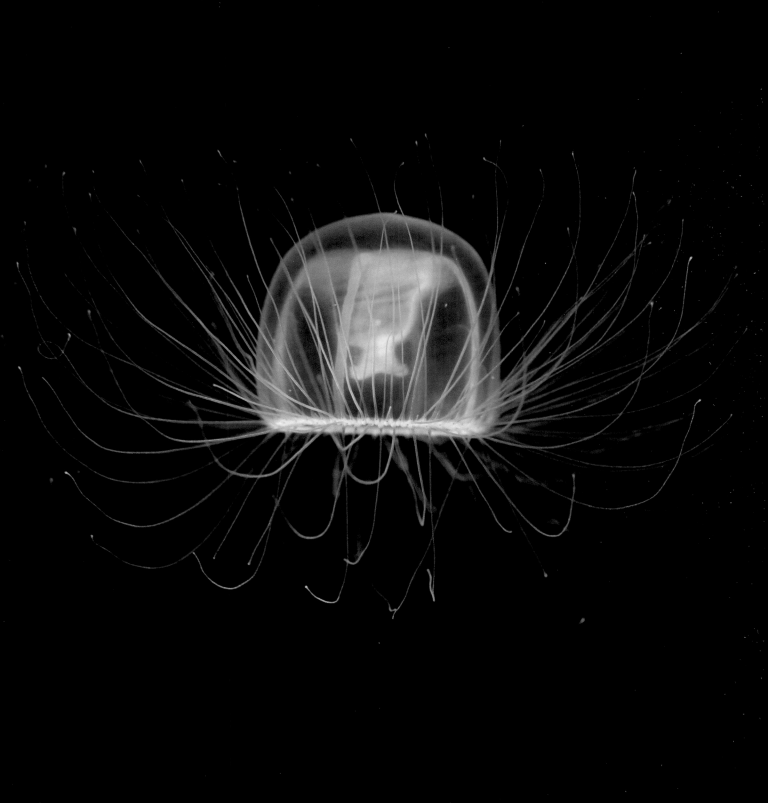

# STALKED JELLYFISH

Stauromedusae have been known for centuries. The trumpet-shaped body; crown of eight arms, each holding a tuft of knobbed tentacles; and slender stalk leading to a sticky foot make them unmistakable. But they are so cryptic that often they are captured only by accident, and they remain understudied.

Stauromedusae are bottom dwellers that lack a pelagic medusa stage (as we discussed in "Anatomy of Benthic Forms," pages 18–19), but that is not the only way they differ from other jellyfish. The staurozoan planula larva is quite distinctive from those of scyphozoans, which are rice-shaped, smooth, ciliated, and free-swimming. In contrast, these appear segmented, with 16 large, hollow, coin-shaped endodermal or internal cells, and are unciliated, moving by a series of creeping extensions and contractions rather than swimming.

**Trumpets of San Juan**

Most of what we know about stauromedusan reproductive biology comes from a small, hard-to-find species from the picturesque, sea-grass-covered coves of San Juan Island in the Puget Sound, where researchers have carried out studies on breeding and development. Surprisingly, however, even to this day, biologists have never been able to raise a stauromedusan through the life cycle.

The San Juan Island species on which the developmental studies were done has been informally known since at least the 1940s; however, it remains unclassified. It may seem hard to believe that a species that is so important to understanding the biology of a whole class of organisms could remain unnamed, but such is the nature of nature. In fact, scientists estimate that less than 15 percent of all species alive today worldwide have been classified so far.

**Scientific name** [all species in the class Staurozoa]

**Phylogeny** PHYLUM Cnidaria / CLASS Staurozoa / ORDERS Cleistocarpida and Eleutherocarpida

**Notable anatomy** trumpet-shaped body with eight clusters of short, knobbed tentacles

**Position in water column** benthic, in neritic zone

**Size** mostly to about 2.5 cm (1 in.) tall

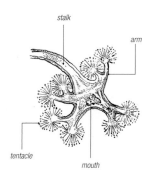

stalk
arm
tentacle
mouth

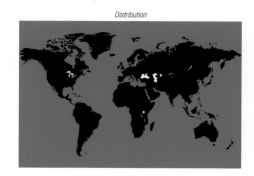

Distribution

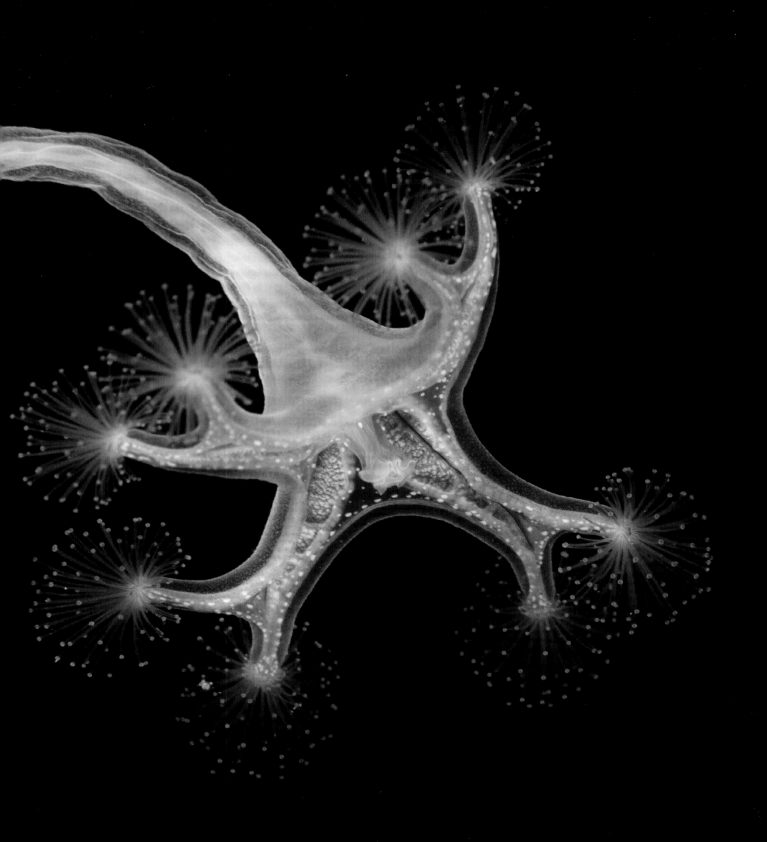

# SAILORS ON THE WIND

Two of the most enchanting jellyfish species known are the By-the-wind Sailor (*Velella velella*) and the Blue Button (*Porpita porpita*). Both live at the air-water interface. Stunningly blue and gracefully traveling the seas one breeze at a time, these creatures are actually upside-down floating polyp colonies. During periods of sustained onshore winds, huge flotillas of these wayfarers may be stranded along coastlines.

The main bulk of the body of these hydrozoans is a disklike chitinous skeleton comprising concentric sealed air chambers, which provide buoyancy. In *Velella* the skeleton is oval and has an oblique sail, whereas in *Porpita* the skeleton is round and absent a sail. They differ too in the structures along the skeleton's margins: in *Velella* many short tentacles hang down like blue icicles, while *Porpita* has numerous long tentacles, knobbed along their length and radiating outward like the arms of a stylized sun.

## Below the Surface

The undersurface holds a teeming colony of polyps, packed together like taste buds on a tongue. They sprout off tiny medusae, structurally simple and each no larger than a grain of sand. These medusae are as delicate as pollen and live only a short time; their sole function is to be dispersal vessels for reproduction.

Polyps of *Velella* and *Porpita* catch their planktonic food with tentacles. While their nematocysts are toxic to their tiny prey, they are harmless to humans. The nudibranchs (sea slugs) that prey on them are able to concentrate these stinging cells in their own tissues and use them for defense.

*Porpita* is generally found in tropical waters, whereas *Velella* is more widely distributed in temperate waters as well.

**Scientific name** *Porpita porpita* and *Velella velella*

**Phylogeny** PHYLUM Cnidaria / CLASS Hydrozoa / ORDER Anthoathecata

**Notable anatomy** flat, chitinous, chambered disk with or without a vertical sail, with polyps underneath

**Position in water column** floats at air-water interface

**Size** skeleton 1–6 cm (0.4–2.35 in.) in diameter

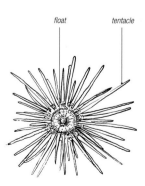

float  tentacle

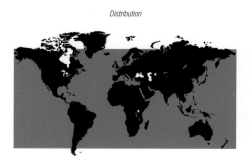

*Distribution*

# PLATYCTENES

People gazing into reef aquariums are often rewarded with the unexpected sight of small, colorful, oval filmy patches gliding over algae or soft corals or of two long feathery tentacles waving in the current. These are benthic ctenophores called platyctenes, common creatures rarely recognized for what they are.

These flatworm-like, bottom-dwelling relatives of the more familiar comb jellies appear most of the time as no more than a thin film, perched on the shaft of urchin spines, in the crotch of a coral branch, or on the ends of algal strands. Every now and then, they puff up the two ends of their oval body into chimneys and unfurl their long, finely side-branched tentacles to catch a meal of plankton or organic particles.

## Life Stages

Platyctenes are hermaphroditic, with the male and female reproductive cells occurring side by side within the body. Embryos are brooded on the underside of the mother, until they develop into a larva that resembles a sea gooseberry (page 44). The larvae swim for a brief time before they settle on a suitable surface, orient themselves mouth down, and then simply open the mouth and flatten out. Their everted pharynx, or throat, becomes their ciliated underside.

Platyctenes are the only members of the phylum Ctenophora to have a benthic life stage, and in contrast to most of the medusae of the phylum Cnidaria, they live on the bottom even in the adult stage. They are found in all marine habitats of the world.

**Scientific name** [all species in the order Platyctenida]

**Phylogeny** PHYLUM Ctenophora / CLASS Tentaculata / ORDER Platyctenida

**Notable anatomy** flat, oval, filmy body with two feathery tentacles

**Position in water column** benthic, in neritic zone, creeping on corals, algae, or sea urchins

**Size** generally less than 2.5 cm (1 in.) long

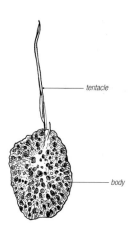

tentacle

body

*Distribution*

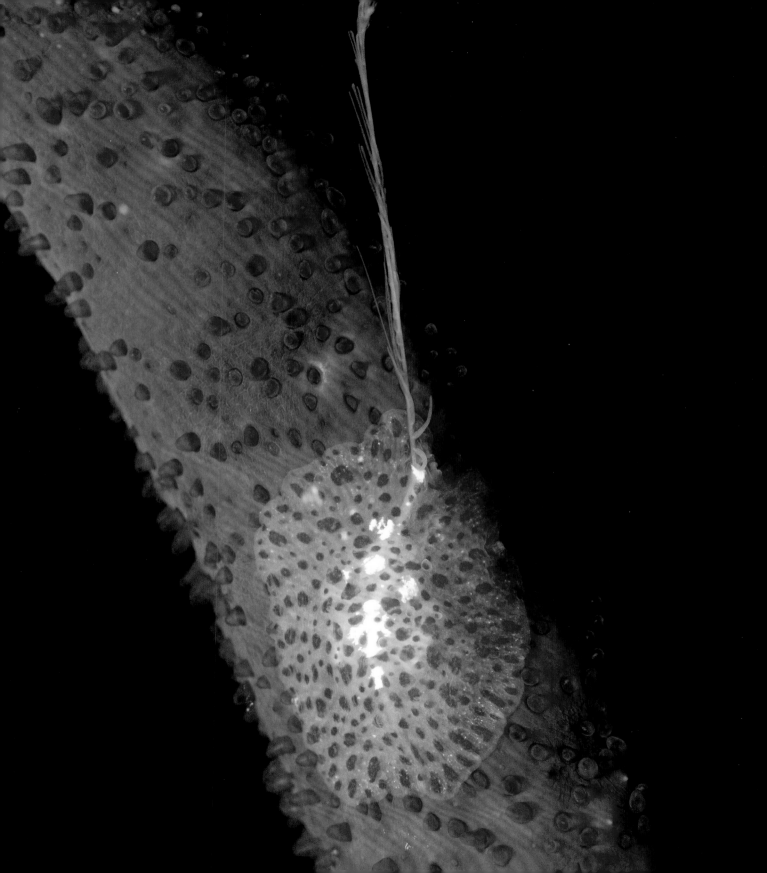

# TADPOLE LARVA

THE APPENDICULARIANS, or larvaceans, are among the most unusual creatures in the ocean. Although they are often called Tadpole Larvae, they aren't actually larvae of tadpoles. The name comes from their appearance: they have a teardrop-shaped body and a broad, flattened tail. Tadpole Larvae are in our phylum, and as such, they have rudimentary forms of organs similar to ours: heart, brain, and so on. Appendicularians are neotenic, that is, retaining larval characteristics in the adult stage. These characteristics include the possession of a notochord, the evolutionary and embryological precursor to the spinal column.

*Oikopleura* and its kin build mucous feeding structures that look like three-dimensional cobwebs, in which food particles are filtered through layers of progressively smaller windows or holes. The animal sits inside its mucous house, driving the filtering process while drifting along by rhythmic undulations of its tail.

## Moving House
They build and discard and rebuild several such houses each day. When the maker of the house moves out, the house collapses but may persist for some time as an amorphous blob. These houses slowly sink in a rain of "marine snow", bringing trapped particles to the deep sea where food is often scarce.

Smaller tadpole larva species like *Oikopleura* and *Fritillaria* are common in surface and coastal waters, whereas *Bathochordaeus* is found in the open ocean below about 80 meters (265 feet).

**Scientific name** *Oikopleura* spp.

**Phylogeny** PHYLUM Chordata / SUBPHYLUM Tunicata / CLASS Appendicularia / ORDER Copelata

**Notable anatomy** tadpole-shaped, with a broad, flat tail, living in a delicate mucous envelope

**Position in water column** epipelagic to mesopelagic, in open ocean

**Size** body typically 1–2 mm (0.04–0.08 in.) long; house typically less than 1 cm (0.4 in.) diameter

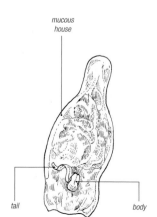

mucous house

tail

body

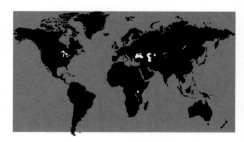

Distribution

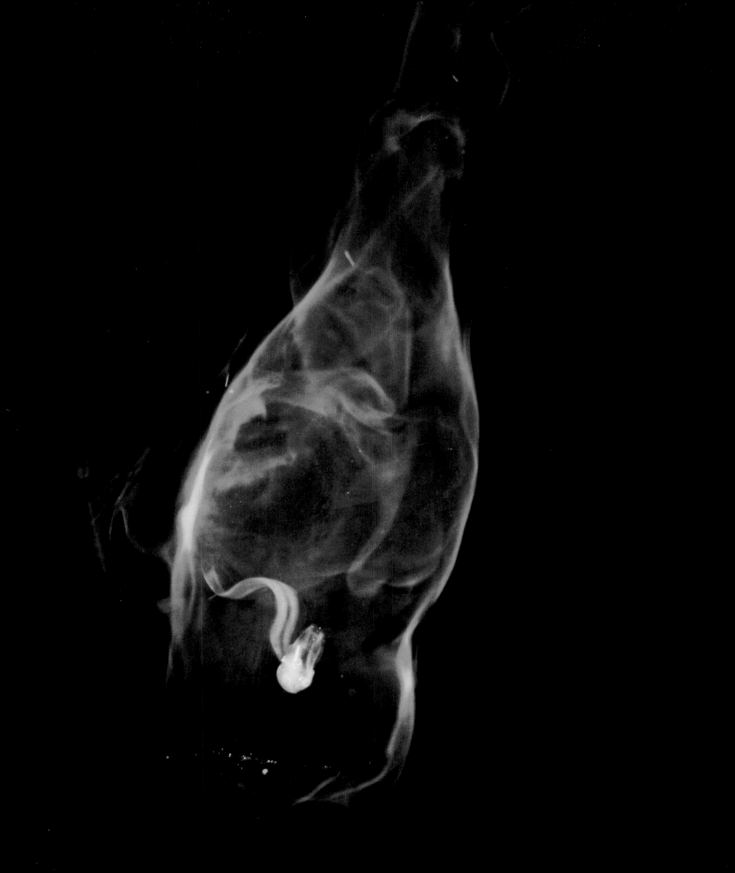

# BATTERING RAM JELLIES

Living in the deep sea is fraught with challenges. Food and mates are hard to find, so it is not uncommon to see strange adaptations in the creatures faced with these difficulties. Another problem is finding suitable surfaces on which to live, especially for those species found in the mid-water—that vast, open nothingness between the sea surface and the bottom. Some jellyfish have evolved to have no polyp stage. Others have evolved a parasitic polyp; the narcomedusae fall into this group of creatures.

## Life Cycle Adaptations

Some narcomedusae develop as parasitic stolons in or on other jellyfish species; young medusae are budded from these stolons. Others develop as parasitic medusae in the stomach of the parent or other species. Still others lack a parasitic stage but simply develop directly from the planula larva straight to a small version of the adult form. And some have a free-swimming intermediate larval form called an actinula, which resembles a stalkless polyp. For most, the life cycle is unknown.

## Tentacles

Narcomedusae are readily identifiable from all other medusae by their tentacles, which are filled with solid cells and resemble a stack of coins. They always attach to the body above the bell margin. The most alien-looking species are in the family Aeginidae with usually two or four tentacles. Like all narcomedusae, aeginids swim with their tentacles held stiffly out forward like battering rams.

**Scientific name** *Aeginopsis laurentii*

**Phylogeny** PHYLUM Cnidaria / CLASS Hydrozoa / ORDER Narcomedusae

**Notable anatomy** small pyramidal body with four solid, antenna-like tentacles

**Position in water column** epipelagic to mesopelagic, in open ocean

**Size** bell up to about 5 cm (2 in.) in diameter, often much smaller

tentacle

gonad

bell

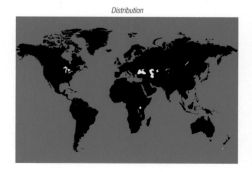

*Distribution*

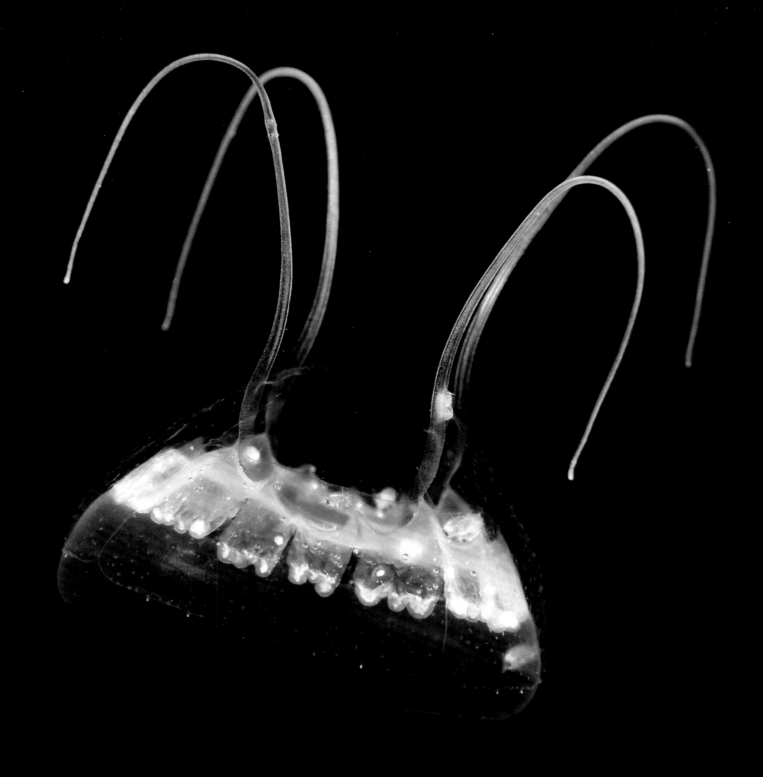

# GORGONS OF THE POND

ONE OF THE STRANGEST of all species types in the broad jellyfish group Medusozoa, or medusa animals, is not a medusa at all. *Hydra* is a genus containing more than 40 species of small, simple, naked hydroids, in which the medusa stage is absent. The asexual proliferation stage and the sexual dispersal stage are combined in the polyp. And each *Hydra* polyp is solitary, not connected to other members of its colony, as in most hydroids.

There is another unusual thing about *Hydra*: it is found exclusively in fresh water. Most other jellyfish and hydroids are marine. Despite these aberrations, *Hydra* has been considered for centuries to be the archetype of the Hydrozoa order of plantlike animals and water jellies.

*Hydra* is common worldwide in freshwater ponds and lakes, where it may be found attached to stones and leaves. The body is long and tubular, with a crown of usually five or six tentacles at the unattached end. Depending on species, it may reach up to about 1.5 centimeters (just over half an inch) long. It reproduces clonally by budding daughter polyps off the column within a specific zone near the base. Sexual reproduction in *Hydra* is more complicated: most species are dioecious (each individual is either male or female), but some are hermaphroditic. Testes and ovaries are typically spread along the external body wall, but may be confined to specific zones.

## Healthy Appetite

*Hydra* eats mainly small crustaceans, worms, and insect larvae, but may also eat newly hatched fish and tadpoles. The stomach can extend immensely to accommodate large food items. At least one species of *Hydra* is estimated to live for 1,400 years.

**Scientific name** *Hydra* spp.

**Phylogeny** PHYLUM Cnidaria / CLASS Hydrozoa / ORDER Anthoathecata

**Notable anatomy** small, slender, naked, solitary polyp with a crown of tentacles

**Position in water column** benthic, or suspended from surface film in fresh water

**Size** to 1.5 cm (0.6 in.) long

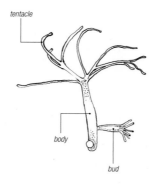

tentacle

body

bud

Distribution

# HERPES JELLYFISH

As farfetched as it seems, a person can have a herpes-like reaction to the sting of *Copula sivickisi*, the Herpes Jellyfish. It does not actually impart the herpes virus, but it leaves a wound that looks and feels similar to a fever blister and recurs periodically for many years. The creature's striking color pattern warns of its virulent sting: the tentacles are banded yellow and brown.

## Bottom Dweller

The diminutive *Copula sivickisi* is remarkable in other ways too. It is more or less benthic at every stage of its life cycle. This tiny, box-shaped species, just eight millimeters (about a third of an inch) tall, is able to swim but spends most of its time resting on the sides of objects, held in place with sticky pads on the top of its body. In this posture, with one or two of its tentacles streaming out in the current, it can passively troll for small fish or plankton.

## Mating

*Copula*'s breeding process is remarkable too. When females are ready to mate, they develop black dots around the margin of the bell. Males perceive these dots and begin a courtship with the females. Eventually the male places a sperm bundle onto the female's tentacles; after ingestion of the sperm, the fertilized eggs are visible as bright orange bodies inside her gastric cavity. Hours later, the female finds a suitable bit of seaweed, where she lays a mucous strand teeming with developing planula larvae.

*Copula* is primarily a warm-water species, commonly found in tropical and subtropical localities of the world, including the Caribbean, Hawaii, Japan, and the Great Barrier Reef of Australia, but has also been found in a few temperate regions, including Tasmania and Wellington, New Zealand.

**Scientific name** *Copula sivickisi*

**Phylogeny** PHYLUM Cnidaria / CLASS Cubozoa / ORDER Carybdeida

**Notable anatomy** small, cube-shaped body, with a single tentacle suspended from each corner

**Position in water column** more or less benthic, in shallow water

**Size** bell to 8 mm (0.3 in.) in height

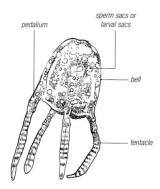

pedalium
sperm sacs or larval sacs
bell
tentacle

Distribution

# PURPLE PEOPLE EATER

THE NAME *Pelagia noctiluca*, Latin for "drifting night light," comes from the brilliant bioluminescence the species displays after dark. Masses of *Pelagia* glowing cool blue-green must have been quite a wondrous sight for sailors in the 1600s and 1700s when the species was discovered.

## Purple Blooms

*Pelagia noctiluca*, known as the Purple People Eater or Mauve Stinger, is a common pest along tropical and subtropical coastlines of the world, especially in and around the Mediterranean. It is a handsome species, with a mauve hemispherical body bearing eight long, fine tentacles and four ruffled oral arms. But it travels en masse, and its sting is fierce. In recent years its blooms have shut down swimming along entire coastlines of Spain and France and have put salmon farms out of business in Ireland.

*Pelagia* has a broad diet and a voracious appetite. Its prey includes hydromedusae, ctenophores, small crustaceans, fish eggs, and other plankton. Its dense swarms can quickly consume just about every living thing in the water.

In addition to being so familiar and fearsome, *Pelagia noctiluca* is also one of the most aberrant and intriguing jellyfish when it comes to life history. It has no polyp form. Whereas in most scyphozoans the planula larva settles to the seafloor and turns into a polyp, which then produces medusae through a process of strobilation, the *Pelagia noctiluca* planula flattens out and transforms directly into an ephyra, or larval medusa. The whole developmental process takes about 92 hours: it spends its first 48 hours as a cilia-powered swimming planula, then it takes 44 hours to metamorphose into a muscle-powered swimming ephyra.

**Scientific name** *Pelagia noctiluca*

**Phylogeny** PHYLUM Cnidaria / CLASS Scyphozoa / ORDER Semaeostomeae

**Notable anatomy** dome-shaped body with eight tentacles and four pleated oral arms; lovely mauve color

**Position in water column** epipelagic, usually in neritic zone, may be found in open ocean

**Size** bell to 6.5 cm (2.6 in.) in diameter

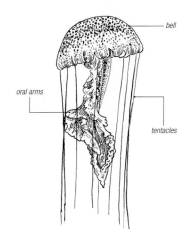

bell

oral arms

tentacles

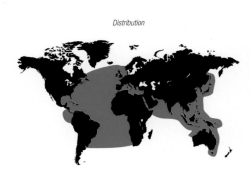

Distribution

# BRACKISH GODDESS

**W**E USUALLY THINK of invasive species as ugly and foreboding. *Maeotias marginata* challenges that stereotype. *Maeotias* is one of the most gloriously, wildly beautiful of all jellyfish. The bell is large for a hydromedusa and almost perfectly hemispherical—resembling half of a ping pong ball in size and shape—with a fringe of hundreds of fine tentacles.

### Leaping Into Action

It sits on the bottom probing the sediments with its long manubrium and nimble lips, looking for food. And every now and then, for reasons known only to the jellyfish, it jumps up and begins pulsing around. Then in the middle of the water or near the surface, it simply stops pulsing and slowly drifts down with its tentacles all deployed. It looks like a firework caught in freeze frame just at the right moment.

### Stowaway

Native to the Ponto-Caspian region (the Black and Caspian Seas), ballast water transported by ships has enabled *Maeotias* to expand its range to faraway places including China, the Mediterranean and Baltic Seas in western Europe, and the Chesapeake and San Francisco Bays in America. Indeed, in the upper reaches of the San Francisco estuary lies a small community called Suisun City. In the shadow of the Jelly Belly jellybean factory, this jellyfish can be found in abundance during the summer months.

Perhaps surprisingly, given its invasive nature and stunning beauty, little is still known about the biology and ecology of *Maeotias*. Intriguingly, however, it is often found as part of a trio of introduced hydrozoans, the others being the less conspicuous *Blackfordia* and the diminutive *Moerisia*.

**Scientific name** *Maeotias marginata*

**Phylogeny** PHYLUM Cnidaria / CLASS Hydrozoa / ORDER Limnomedusae

**Notable anatomy** hemispherical, with many fine tentacles around the bell margin

**Position in water column** shallow water in enclosed brackish bays

**Size** bell to about 5 cm (2 in.) in diameter

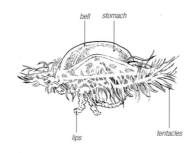

bell    stomach

lips    tentacles

*Distribution*

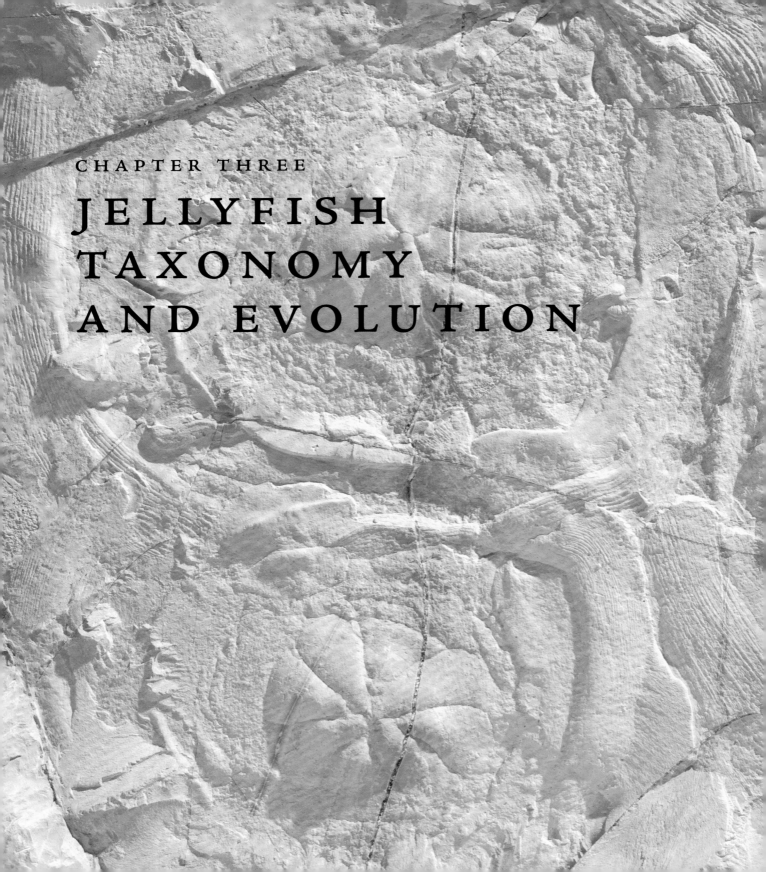

## CHAPTER THREE

# JELLYFISH TAXONOMY AND EVOLUTION

# INTRODUCTION TO TAXONOMY AND EVOLUTION

In this chapter, we look at jellyfish diversity through time. This subject encompasses two main fields of study, which look into what the creatures are, and how they are related to one another. Taxonomists, scientists who examine the identification and classification of species, cover the first of these areas. Systematists or phylogeneticists cover the second, determining how species are related by using DNA and other comparative analytical tools to understand the relationships between different taxa (singular, *taxon*, a unit of classification such as species, genus, or family). Researchers use branching diagrams called cladograms or phylogenetic trees (similar to family trees) to pictorially represent hypothesized relationships between species; this field of study is known as cladistics.

### The Gelatinous, Transparent Form

The creatures that we collectively call "jellyfish" are not as alike as they may seem, even though all are gelatinous, typically transparent, drifting animals. Jellyfish are far more diverse in their identities, as well as in their biology and ecology, than birds and frogs, or humans and lizards, or insects and lobsters. Their transparent, gelatinous nature has evolved through adaptation to a similar environment rather than through shared ancestry.

Transparency is thought to have evolved as a means of camouflage. It is also possible that it evolved as an energetically less expensive alternative to pigments, in an environment where pigment and associated camouflage are not important. The open ocean does not have nooks and crannies for creatures to try to blend into, especially in the depths beyond where light penetrates. Other oceanic creatures unrelated to jellyfish have also evolved transparency, including crustaceans, squids, worms, and even fish.

The gelatinous body is thought to have evolved as a means of both buoyancy and protection, and is probably also less costly for these animals to produce in their energy-limited habitat. Heavy bones or meaty flesh would require extra energy input to keep afloat, whereas jellyfish are nearly neutrally buoyant and very energy efficient. Also, because gelatinous material costs the animal less, in terms of energy, to make than flesh or bones, the animal can make more tissue without expending much energy, thus attaining a larger size and becoming more resistant to predation. A gelatinous body therefore offers a way for jellyfish not only to survive in their sparse, drifting world but to dominate it.

### The Importance of Scientific Studies

We also look in this chapter at the paleontological and developmental history of jellyfish. Fossils of these creatures—though surprising to consider—actually form a fascinating body of evidence for our understanding and interpretation of their evolutionary development. And jellyfish development offers comparative insights into our own evolution, such as how our brains develop and how we as organisms have evolved from our ancient ancestors.

As we learn more about the species of jellyfish and their roles in ocean dynamics, it is becoming increasingly clear that taxonomy and phylogeny are the basis for all other types of study, including biology, ecology, and management of blooms and stings. Taxonomy allows us to reliably and consistently identify and manage problem species, as well as communicate meaningfully about them, whereas phylogeny allows us to predict features and behaviors in species we do not know, based on their relationships to species we know more about.

Taxonomy also offers a unique way for scientists to connect with the general public. Scientists feel a very special connection to new species (they are exciting!), and through good identification tools can provide nonscientists with a means of experiencing nature and its inhabitants in a familiar and meaningful way. This is not entirely a public service

**Below** Studies of taxonomy and phylogeny help us understand and predict complex features of biology and ecology: for example, the seasonality of *Carybdea* or relative severity of its sting.

performed by scientists; people acting as citizen-scientists often serve as the eyes and ears for wondrous discoveries.

Another, perhaps more fundamentally important thing is becoming clearer as we learn more about jellyfish: how little we actually know about these strange and marvelous beasts. In the last 20 years alone, hundreds of new species and relationships have been revealed, and some of these discoveries have been jaw-droppingly amazing: New species. New families. New orders. The largest invertebrate discovered in the twentieth century (Black Sea Nettle, *Chrysaora achlyos*, page 114.) Lower taxa have been elevated to higher. Life cycle studies have united some disparate taxa, while other taxa have been moved around in the evolutionary tree. The number of known jellyfish species whose stings are potentially lethal to humans has grown from three to 20. Yet we have only just scratched the surface in understanding jellyfish and their strange world, and incredible opportunities for discovery no doubt await curious researchers.

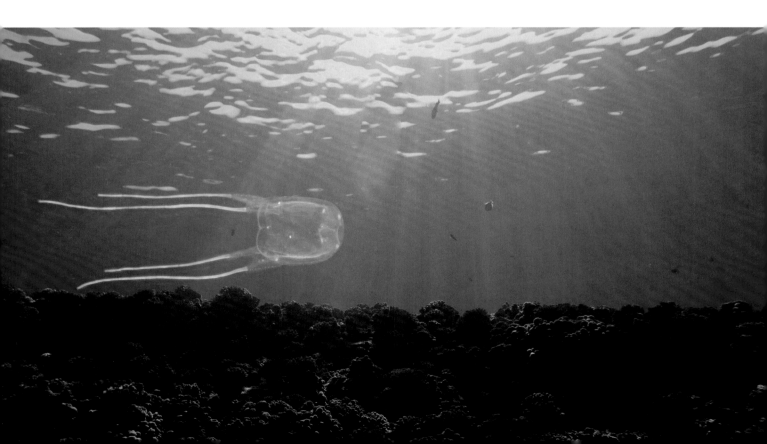

# JELLYFISH IN THE FOSSIL RECORD

When we think of fossils, we may immediately conjure thoughts of big, heavy dinosaur bones or petrified tree trunks. It is hard to imagine how soft-bodied creatures such as jellyfish could possibly fossilize. But fossilize they do, under the right circumstances. Jellyfish fossils do not leave a body in the same way bones do; rather, they leave a footprint of sorts, a shallow impression.

INTERPRETING JELLYFISH in the fossil record relies on the presence of certain characters in their otherwise simple bodies. Radial symmetry is a strong indicator, as sometimes the mouth or the reproductive organs may fossilize. Concentric symmetry is also indicative, although sometimes it may be derived from other types of animals or plants. Sometimes tentacles or comb rows may fossilize in ultrafine sediments. Surprisingly, however, no salp fossils have yet been found.

### Jellyfish of the Precambrian Era

The earliest accepted jellyfish fossils come from the Ediacaran, the latest period of the Precambrian era, which lasted until about 585 million years ago. The period is named for the location in which these fossils were first discovered, the Ediacara Hills in Australia. Deposits of similar fossils have since been discovered along the White Sea of Russia, the coastal bluffs of Newfoundland, and the hills of Namibia. Earlier fossils are still debated (page 126).

When the Ediacaran fossils were discovered in 1946, many of them were originally interpreted as jellyfish. Several have since been reinterpreted as other forms of life. A few, however, remain as splendid examples of early jellyfish.

**Below** Fossil interpretation is not always straightforward. This stunning fossil, *Paleophragmodictya*, is classified as a sponge. It shares characteristics with sponges and medusae, suggesting that we have only scratched the surface on understanding early animal evolution.

Two species in particular are worthy of mention. *Conomedusites lobatus* is a small form that strongly resembles the coronate medusae alive today, such as *Periphylla* and *Atolla* (pages 166 and 168). *Albumares brunsae* is another small fossil form, this one similar to the common Moon Jellyfish (*Aurelia aurita*, page 130), complete with nearly identical radial canal branching; however, *Albumares* is based on a three-parted symmetry rather than four.

### Jellyfish of the Cambrian Era

The Precambrian era terminated with an abrupt shift, an evolutionary "Big Bang" known as the Cambrian explosion. In a geological short time later in the Late Cambrian, blooms of jellyfish were stranding and fossilizing in repeated bedding planes in what are now Wisconsin sandstones.

In the 1980s, a stunning discovery was made in Yunnan Province in southern China—the Chengjiang shale formation, a deposit from the Early to Middle Cambrian period, about 520 million years ago. This extraordinary fossil site is known for the most meticulously detailed soft-bodied organisms yet discovered. Although ctenophores have been found occasionally in other deposits, three species have been found here so far. Several other *Lagerstätten*, or fossil sites with exceptional preservation, also contain jellyfish.

### Jellyfish of the More Recent Past

More than 200 million years after the Cambrian explosion, box jellyfish and other soft-bodied organisms were fossilized in fine ironstone sediments. The Mazon Creek formation in what is now northern Illinois contains species dating back

**Above left and above** A sandstone quarry in Wisconsin (above left) is home to an unnamed and unclassified species of fossil medusa (above right). Thousands stranded in seven consecutive bedding planes, providing a unique glimpse into ancient jellyfish blooms. As their muscles excavated the sand in an attempt to escape, they built a characteristic pattern of a filled mouth surrounded by concentric rings that allows us to identify them some 500 million years later.

to the mid-Pennsylvanian epoch of the Carboniferous period, some 300 million years ago. These medusae fossilized in such exquisite detail that we can perceive many of their structures; they would be considered no different to any other medusae if found alive today.

The Montral-Alcover formation in Catalonia, in northeastern Spain, is another deposit of exceptionally well-preserved fossils, this one from the late Ladinian age of the Middle to Upper Triassic, about 235 million years ago. Two of the species bear a striking resemblance to living forms of hydromedusae (for example, *Aequorea victoria*, page 198), while a third has no modern counterpart but has been interpreted as a jellyfish.

Younger still is the Solnhöfen limestone formation in Bavaria, Germany, most famous as the site where the earliest known bird fossil, *Archaeopteryx*, was discovered. However, the site also contains several species of jellyfish, including large, well-formed fossils strongly resembling today's blubbers. These jellyfish were entombed some 155 million years ago during the Late Jurassic period.

Intriguingly, all these forms are nearly indistinguishable from jellyfish species living today, suggesting that jellyfish as a successful life form have not changed much in nearly 600 million years.

# SYMMETRY VARIATION

One of the fundamental characteristics of the Medusozoa is that its members' bodies are based on a tetraradial (four-parted) plan. However, as with so many things in nature, there can be exceptions to rules. Some species and families are based on hexaradial (six-parted) or octoradial (eight-parted) forms as the norm. Even in normally tetraradial species, about 2 percent of the individuals in most wild populations will vary in the expression of their symmetry, generally anywhere from one- to eight-parted. One of the very strange things about jellyfish, however, is that their clones are not always identical: the free-swimming "individuals" that we observe are clone mates of other "individuals," so with these varying expressions, the concept of an identical clone can be obscured.

LABORATORY REARING EXPERIMENTS have demonstrated that a single jellyfish polyp can bud off medusae of different symmetries. The symmetry of a given medusa is set during its development, but it is not clonally consistent. A single polyp can therefore give rise to tetraradial, pentaradial, and hexaradial medusae, all at the same time. These clone mates are genetically identical but far from identical morphologically. They look so different, in fact, that symmetrically variant individuals have sometimes been classified mistakenly as different species. Even more intriguing, as some polyps continue to bud new polyps and medusae over time, their symmetry expression becomes increasingly unstable, and they release a higher percentage of non-tetraradial medusae over successive cloning events.

We generally think of cloning as keeping things constant, while sexual reproduction is where the variation occurs. However, in the medusae it works the other way round. The increasingly unstable expressions of symmetry are reset back to a more typical ratio when these jellies reproduce sexually. The genetics underlying this phenomenon are unclear.

**Right** Jellyfish clone-mates may be different symmetries and sexes. In the moon jellyfish (*Aurelia* spp.), for example, a single clonal event of strobilation, a special kind of budding, may produce symmetries from one to eight, like the three, four, and seven shown here.

## Symmetry and Survival

Variation in symmetry also has potentially intriguing evolutionary implications. In wild populations, the most common non-tetraradial symmetry expression is six-parted. The next most common are five-parted and then three-parted. Many Precambrian radially symmetrical organisms were three-parted. One possible reason for this has to do with oxygen levels.

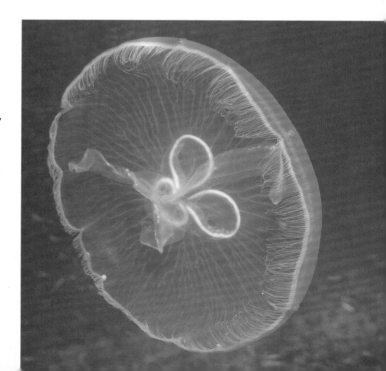

Oxygen—both atmospheric and dissolved in the sea—was present at much lower levels in the late Precambrian era than it is today. Living organisms, of course, need oxygen to drive their cellular processes. Studies of jellyfish may offer insight into the ecology of early symmetry expression. Laboratory observations have shown that jellyfish with higher symmetries pulsate more frequently than those with lower symmetries. In other words, a medusa with six stomachs and 12 sense organs (rhopalia) will pulsate considerably more often and use up considerably more energy than a medusa with four stomachs and eight sense organs. Expending more energy requires uptake of more oxygen, and in an environment with limited oxygen, this need could mean death.

There is a trade-off with reduced symmetry, though. Medusae with higher symmetries also have more reproductive organs. Even though each of the gonads may be about the same size, a medusa with six has half again as many as a medusa with just four. More gonads means more sperm or eggs, which, all other things being equal, takes more energy to produce but means a higher likelihood of leaving great numbers of offspring.

Today oxygen is not such a problem, but food is. In a food-limited environment, a medusa expending too much energy is likely to starve to death, while its less energetic brethren will survive. But in an environment with plentiful food, the medusa with greater reproductive capability has the evolutionary advantage over those with less.

One of the major challenges to a species' survival is the ability to cope with a variety of ecological circumstances that may arise in the future. Genetic variation that underlies slight morphological or behavioral differences may favor one individual over another in the struggle for survival. And those individuals who survive leave offspring.

It seems that jellyfish have solved this dilemma of an uncertain future—whether the uncertainty may be oxygen or food supply—by budding off a range of symmetries. Because the clone mates from a polyp are genetically identical, it really does not matter which one leaves offspring, only that someone does.

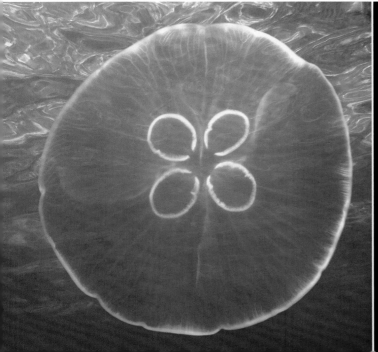

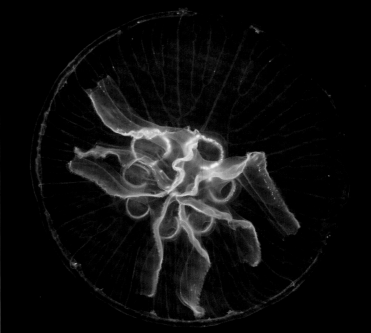

# EVOLUTIONARY RELATIONSHIPS

In the scientific quest to understand evolution, jellyfish provide a fundamental link to early animal life. By understanding the internal and external relationships of jellyfish groups, we can better resolve some of the outstanding evolutionary questions. One of these questions is centered on evolution in the phylum Cnidaria. Which came first, the polyp or the medusa? This question has been debated for well over a hundred years and has only begun to be resolvable with DNA.

### Relationships in the Cnidaria

For a long time the Cnidaria was construed to include three evolutionarily equal classes: Anthozoa, the anemones and corals; Hydrozoa, the water jellies and siphonophores; and Scyphozoa, the true jellyfish. Anthozoans spend their whole life as a polyp, while the medusozoan classes Hydrozoa and Scyphozoa typically have polyp and medusa stages. Some researchers argued that the fact that medusae have more complex nervous systems than corals and anemones provides evidence that the Anthozoa was the more ancestral state. Other authors countered that most medusozoans have a large medusa body and a small polyp, and tend to be simpler in general morphology than anthozoans, and that this provides evidence that the medusa was more ancestral. Recent genomic studies have lent support to the anthozoan-first hypothesis, putting the polyp before the medusa.

Much of this evolutionary insight has come about through the clarification of the relationships within these previous three classes. For most of the past couple hundred years of scientific history, the Scyphozoa was interpreted to include all the medusa groups that were not in the Hydrozoa. The box jellies had been in the Scyphozoa group, but in 1975 they were pulled out and placed their own class, the Cubozoa, because of their unique polyp characteristics and metamorphosis. The stauromedusae were in the

Scyphozoa too, but in 2004 were put into their own group, the Staurozoa, after analysis of their morphology and DNA. More recently, it has been suggested that the coronate medusae (order Coronatae, currently in the Scyphozoa) should be pulled out into their own group as well.

This leaves only the Semaeostomeae (moon jellies, sea nettles, and their kin) and the Rhizostomeae (blubber jellies) in the Scyphozoa. However, the classification of the latter of these, too, is tenuous. The two main divisions within the Rhizostomeae now appear to be an artificial grouping. They look similar, but they are not, in fact, each other's closest relative. The grouping is polyphyletic (*poly* = multiple, *phyletic* = evolutionary origin). Other, more familiar, examples of polyphyletic groups are flying things (birds, bats, and insects) and even the grouping "jellyfish" (which includes members of three otherwise unrelated phyla).

### Relationships among Jellyfish and Other Life Forms

Another long-standing evolutionary question is the branching order of the lower animal groups such as cnidarians, ctenophorans, sponges, and worms. By historical convention, the tree of animal life started with sponges at the base, and then branched off into cnidarians, then ctenophores, then worms, and so on, as we worked our way up the tree. Humans were interpreted to be at the very top of the tree, the pinnacle of evolution.

Different DNA studies, however, have placed the Ctenophora in different places relative to the Cnidaria. Some studies have placed the ctenophores with the cnidarians as the sister group to the Bilateria (animals with bilateral symmetry; i.e., all higher from worms to humans). Other studies have grouped the ctenophores with the Bilateria, branching off from the cnidarians. One recent study

suggested that the Ctenophora was ancestral to the Cnidaria, which is contrary to interpretations that consider the more sophisticated structures and development of the former relative to the latter. This highlights the difficulties of trying to interpret a whole picture of something from mere fragments: one gene may give a very different result compared to another gene. Further studies on whole genomes are likely to answer these questions more satisfactorily.

The relationships of classes, orders, and families within the phyla are being reinterpreted as well. The classical notion that all families in an order, for example, are hierarchically equal is being dispelled. Comparison of DNA allows us to test which groups are ancestral to others. As with higher-level studies, this is not without its complications: one gene often gives an entirely different answer than another gene. However, each study gets us closer to ultimately understanding the relationships of the organisms with which we share this great rock.

**Left** Ernst Haeckel, who created the original sketch for this lithograph of jellyfish medusae, famously developed the idea that ontogeny recapitulates phylogeny, meaning that as an organism progresses through its developmental stages, so too it progresses through the forms of its ancestors.

**Above right** The great exploratory expeditions of the nineteenth century produced an enormous amount of new species and new understanding. These men aboard the HMS *Challenger* collected jellyfish for the taxonomist and evolutionary philosopher Ernst Haeckel.

# EVOLUTIONARY PATTERNS IN SALPS AND THEIR KIN

Even though we include salps, or pelagic tunicates, and some of their close relatives within the category "jellyfish," this is a completely unnatural, or polyphyletic, grouping. The species we recognize collectively as jellyfish evolved to look like each other through similar lifestyles, not through shared ancestry. Like other more familiar jellies such as medusae, ctenophores, and siphonophores, salps are typically transparent and have soft, gelatinous bodies. But the similarities end there.

## Phylum Chordata

Salps, as we have noted, are in our own phylum (Chordata), so they are more closely related to humans than to the other jellyfish groups. We share key embryonic features, such as possession of a notochord, the evolutionary and developmental precursor to the backbone. Their subphylum, known as Tunicata because of the tunic-like cellulose covering of their bodies, contains the appendicularians, ascidians (also called sea squirts), doliolids, salps, and pyrosomes. The phylum Tunicata appears to be the sister taxon to the Vertebrata (animals with backbones), which includes humans. This means the tunicate and vertebrate lineages appear to be each other's most recent common ancestor. However, seeing salps merely as our ancestors or cousins is a very simplistic view of this most complex and intriguing group.

## Relationships in the Tunicata

While the sessile (attached, rather than free-floating) ascidians first appear in the fossil record in the Early Cambrian, fossil examples of the conspicuous pelagic forms—the salps and pyrosomes—have not been found. This is particularly puzzling because salps and pyrosomes tend to have a tougher, more robust body than medusae, and yet numerous medusa fossils are known. Even more surprising, small and delicate appendicularians (also called larvaceans) have been found fossilized in Early Cambrian shales in China.

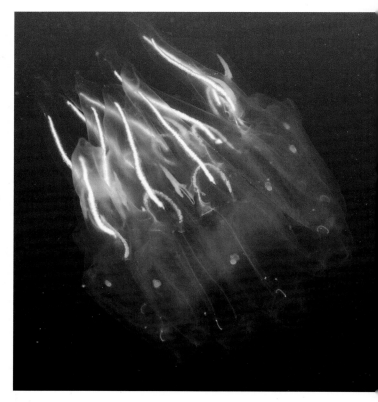

**Above** It may be hard to imagine looking at salps that they are more closely related to us than to other jellies, but their embryological features, tiny brain, and primitive heart give them away.

Appendicularians, or tadpole tunicates (page 82), are an evolutionary anomaly. As their name might suggest, they are tadpole-shaped. In fact, mature appendicularians have a similar appearance to the larvae of most other tunicates. This retention of larval characteristics in the adult form is called neoteny.

Recent DNA analysis indicates that appendicularians are the ancestral form in the Tunicata, and the sister group to all other tunicates. Moving up the tree, ascidians are ancestral to the colonial pelagic forms such as the true salps, the pyrosomes, and the doliolids. The true salps appear to be the most highly derived (evolutionarily advanced) of all the tunicates.

**Above** The aggregate stage of the salp *Pegea* forms long chains that curl into a spiral, in contrast to that of the cyclosalps' wheel-like colony (above left). Most other species of salps form simpler, straight chains.

## Salp Colony Forms

Another curious pattern in salp evolution is the division between two quite different colony growth forms: the linear salps (page 208) and the cyclosalps (page 128). The majority of salp species form linear chains in their colonial aggregate stage. In some of these the zooids, or colony members, are arranged side to side, while in other species they are arranged in a more nose-to-tail or herringbone-like fashion, with the zooids branching off on both sides of a central axis. Cyclosalps, by contrast, form rings with the zooids attached together by long stalks or stems that meet at the center, like the spokes of a wheel.

These different forms of salps are easy to tell apart even when the zooids are separated, as both solitary and aggregate forms of linear salps have conspicuous spherical or globular gut masses located toward the posterior end of the body, while in the cyclosalps the gut mass is either linear, running the length of the body (in the solitary stage) or horseshoe-shaped toward the posterior (in the aggregate stage). Recent studies comparing the DNA of these groups suggest that the cyclosalps branched off as a group from the other salps, which means they are a more derived form that evolved from the other salps.

## Salp Research

Because of their phylogenetic position as the most recent common invertebrate ancestor to the vertebrates, salps and their kin have become of great interest to both evolutionary and developmental biologists. Investigations into the formation of important structures such as our brains and our backbones, and into key developmental processes such as embryology, can help us better understand how we came to be who we are. Salps' brains are not as rudimentary as one might think. The salp brain appears to be but a small ganglion, or mass of nerve cells; however, it is symmetrical and shows a significant extent of differentiation into regions that resemble, to a degree, the central nervous system structures of other chordates. Developmental studies on these creatures, therefore, have particular relevance to our own evolution.

# SPECIES CONCEPTS IN JELLYFISH

Generally we think of a species as a population of similar-looking organisms that naturally interbreed but are not able to breed successfully with others outside their group. This concept may work well for organisms like birds and mammals with a lot of features and readily observable breeding habits. But jellyfish challenge these notions. Jellyfish are simple creatures with few characteristics, and with such ancient lineages, the potential for cryptic diversity is very high. How to make sense of groups with comparatively little information is a question taxonomists have been asking for hundreds of years.

A RANGE OF SPECIES CONCEPTS has been defined for separating and recognizing species. Often a framework established in one group fails in another. The interbreeding approach (above) is called the Biological Species Concept (BSC), and is rarely used in lower animals and plants where hybridization is common. More frequently, the Morphological Species Concept (MSC) is the standard for differentiating most invertebrates.

**The Morphological Species Concept**
Under the MSC, species identification and recognition criteria are based on the structural features of organisms. The emphasis on different characters changes over time (see "Taxonomic Trends," pages 110–11), but the unifying concept is that identification is based on observable physical traits. A major advantage of the MSC is that it is simple to do without expensive equipment or lucky timing.

The MSC, traditionally the only tool available to taxonomists, has grown from its historical practice. During the eighteenth and nineteenth centuries it was common to regard species as having "essences," where they were defined by one or two key features. Any specimens that shared the same essence were regarded as the same species, while those lacking it were considered distinct. The essence was often based on something that today we regard as a genus-level or family-level character, or even something now considered to be without taxonomic meaning. Today, the MSC is still used, but the focus has broadened to a consideration of all the features of the whole animal.

There is a profound limitation inherent to the MSC, however, especially with simple creatures such as many of the jellyfish. A great deal of genetic and ecological diversity may be overlooked in species with few physical characters. Many of the species treated in this book are likely to be far more diverse than previously realized.

## The Genetic Species Concept

Genetics have been used for the last several decades as a powerful tool for both interpreting evolutionary relationships between species and helping illuminate

cryptic species. The Genetic Species Concept (GSC), which separates species based on their genetic makeup, might seem to be the obvious answer to resolving hidden or cryptic diversity. However, for many jellyfish genera and families, we simply do not know enough yet to be able to draw any meaningful conclusions. The hydrozoan genus *Obelia* (page 204) contains about 100 described species, but one late-twentieth-century worker considered only four to be valid, because he was unable to tell the rest apart. This is an obvious case in which genetics may help bring resolution. However, the DNA remains unavailable for more than 90 percent of all jellyfish species, including *Obelia*; obtaining it requires great expense in the collecting and sequencing effort.

There has been a recent push to establish a system of "barcoding" for species identification using short genetic markers. Like the scanner at the supermarket that detects the unique identifier for each product, so too species barcoding allows rapid genetic identification of species. Barcoding may be particularly helpful where morphology fails, such as elucidating the species in stomach contents and fecal samples, and for identifying simple species like jellyfish. It fails, however, if the species detected are not yet barcoded with known identities.

There is another, more philosophical, restrictive side of the GSC. The utility of conceptualizing a species is that once described, it can subsequently be identified by others. Consider the experience of walking in a forest and looking around; we experience a pleasant feeling of familiarity in seeing the pine trees, ferns, sparrows, wrens, and other living things we know. Whether we identify these organisms by common names or scientific names, they are familiar to us as recognizable entities. Most of us would quickly lose interest if a walk in the woods necessitated sequencing DNA to identify organisms. Genetic identification is beyond the reach of most people, even most scientists, so it is less operationally useful.

**Left** Historically, different forms of moon jellyfish were all considered the same species. *Aurelia aurita* (far left) from Europe is the simplest in form and oldest known. *Aurelia labiata* (center) from the Pacific coast of North America differs in having a large fleshy manubrium (throat) and finer scalloping of the bell margin. *Aurelia limbata* (left) from the Arctic has an unmistakable brown pigment band around the bell margin.

# LIFE STAGES AND TAXONOMY

Jellyfish in the phyla Cnidaria and Ctenophora have both pelagic (drifting) and benthic (attached, bottom-dwelling) forms. In some cases these forms are different species, but in many cases within the Cnidaria, these are different life stages of the same species. Although this may seem completely foreign, a familiar group for comparison is the butterflies, which fly, while in their caterpillar form they crawl. This dual benthic/pelagic existence makes jellyfish both interesting and challenging to work with.

**Classification of Polyps and Medusae**

From the beginning of scientific history until only very recently, cnidarian polyps and medusae were classified separately, despite being forms of the same species. In the Hydrozoa, two completely different classification systems were used: the hydromedusae were classified in five orders, whereas hydroids were classified in a system of only two. Moreover, experts worked on either benthic forms or pelagic forms, but rarely both. Although these two systems have been merged in recent years, both conventional and unconventional practices that are grandfathered in have left us with some anomalies.

For example, by convention, zoologists strive to classify only adult (sexually mature) forms. This makes sense, because many species have numerous larval stages that often do not resemble each other or the adult form. However, zoologists

also have a priority rule, which states that the older name is the valid one if it is found that two names apply to the same species. Unfortunately, with polyps and medusae, this sometimes means that well-established medusa names are being changed in favor of lesser-known but older polyp names, causing confusion.

Another unusual practice is the intentional establishment of "bucket" taxa, in which polyps or young medusae without resolved life cycles are placed, pending resolution of their mature identity. The establishment of a taxonomic identity is meant to be based on something real and enduring. This practice of place holding, however, is the taxonomic equivalent of a kitchen junk drawer.

## Management of Polyps and Medusae

The functional division between polyps and medusae is also reflected in their management. For example, almost without exception, our knowledge, data, and current research about jellyfish blooms are focused on the medusae. We look into how many medusae are occurring in a given time and place, what problems medusae are causing, and how to eradicate

the medusae. Certainly, it is the medusa stage that we observe when a crisis occurs. However, it is increasingly becoming evident that it is the polyp stage that acts as the seed bank for the bloom as well as being the perennial component after the short-lived medusae die off.

Understanding jelly species requires understanding all stages; similar to the proverbial chicken and egg problem, attempting to decouple the medusa and polyp proves unhelpful. However, while the medusa may be analogous to the chicken, the polyps find their equivalent in multitudes of eggs, not "egg" in the singular. A tiny fragment of a hydroid (a hydrozoan polyp) can seed a whole new colony. Each polyp can also generate many medusae. A single medusa, meanwhile, can release many thousands of offspring, and can drift to a new location to broaden the footprint of the bloom. Indeed, each successive bloom in an area has the capacity to expand its geographical coverage.

But there is an intriguing side to the contribution of polyps to blooms. Because of polyps, jellyfish may well be the ultimate perfect fish—in the harvestable sense. When a fish or a turtle or a whale is taken out of an ecosystem, whether by natural causes or by fishing, that is the end of it. All of its reproductive capacity—its entire lineage of potential offspring—is finished. With these vertebrate species, there is a one-to-one relationship between bodies extracted from the sea and capacity eliminated. But this is not the case with jellyfish. Because a medusa comes from a polyp, and the polyp persists—budding off more genetically identical medusae, as well as more identical polyps, which in turn bud off even more medusae—extraction of each medusa does not have a one-to-one relationship with loss of reproductive capacity. It is not their breeding that makes them a renewable resource; it is their cloning. We will explore more about jellyfish blooms in chapters four and five.

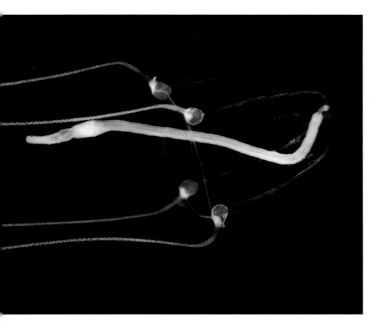

**Left** Many species of jellyfish have very different benthic and pelagic stages and were previously classified in different species, families, and orders. *Sarsia* is one exquisite example. The polyps (far left) form dense thickets on coastal rocks and man-made structures, while the medusae (near left) form ephemeral blooms. In recent decades, there has been a push to resolve the life cycles to link the polyps and medusae.

# TAXONOMIC TRENDS

Two hundred and fifty years ago only 15 species of jellyfish were recognized. These were grouped into two genera: *Medusa*, to which were assigned 11 species of medusae and ctenophores, and *Holothuria*, which held the Portuguese Man-of-war and three species of salps. These were all considered to be mollusks. Today, we recognize about 4,000 species of medusozoans (classes Scyphozoa, Cubozoa, Staurozoa, and Hydrozoa, including the siphonophores), along with about 100 ctenophores and 110 salps and their kin. These creatures span three phyla, eight classes, and 27 orders.

## Historical Trends

By convention, all zoological taxonomy began in 1758. The Swedish botanist, physician, and zoologist Carl von Linné—who Latinized his own name to Carolus Linnaeus and reputedly described *Homo sapiens* based on himself—created the taxonomic system of binomial nomenclature that we still use today. Before this practice of assigning two names, genus and species, to each organism, names were descriptive and unwieldy. For example, the common Mediterranean box jellyfish now called *Carybdea marsupialis* was known as *Urtica soluta marsupium referens*. Some names were even longer. The path from 15 species to thousands has taken many twists and turns.

The century and a half following Linnaeus was the time of the great expeditions in search of travel routes to forge, lands to claim, and riches to take. It was also a time of taxonomic multiplication. Thousands of species were described, many of which have not been found again and remain unrecognized. Some are no longer accepted as valid species. Many species were described repeatedly. Part of the reason for the overlap was lack of communication, as someone in one part of the world would classify a species without knowing it had already been classified by someone elsewhere.

**Below** Many of today's nearly 4,000 jellyfish species bear little resemblance. The class Hydrozoa is one example. *Aequorea* (below) is disk-shaped with gonads on its circulatory system. *Turritopsis* (center left) is thimble-shaped with gonads on its stomach. *Porpita* (center right) is a polyp colony under a floating keratinized disk. *Solmundella* (far right) has gonads in pouches and two bug-like tentacles.

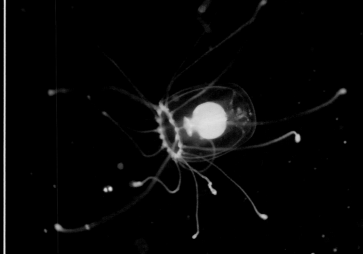

Another complicating factor during this age of discovery was with the criteria for species recognition and definition. The question "What is a species?" may seem straightforward, but history has proven otherwise (see "Species Concepts in Jellyfish," pages 106–7). For most of the past two and a half centuries, the focus was on species essences, a distilled approach that often overlooked true diversity. Although we use different criteria today, this foundation gave us many of the species we now know so well, while many more are only just coming to light.

There was another valuable side to the post-Linnaean period's level of interest and attention: meticulous observations led to hundreds of volumes of published information. Even the smallest aspects of anatomy were studied. Physiology was investigated. Reproduction and growth were observed in detail. Much of what we know today about the biology of jellyfish—and that of so many other creatures too—comes from this era of careful study. Sadly, researchers rarely take the time, have the academic freedom, or can get the funding to make these sorts of detailed observations on species or their biology any longer.

In the early twentieth century, the trend swung around the other way, and many species were merged on the basis of possession of certain key characters. The key characters were those defined by experts. This was also a time of clearing out many of the earlier named species that were unrecognizable due to insufficient or ambiguous descriptions.

## Modern Jellyfish Taxonomic Study

The mid-twentieth century marked the beginning of the modern era of jellyfish taxonomy. There is a push now to recognize species defined on a totality of their characters, and some characters are weighted more heavily than others. For example, in some jellyfish groups the number of tentacles is diagnostic, whereas in others the number varies considerably and thus is not a reliable character for identification. Similarly, the shape of reproductive organs is generally considered to be of high taxonomic value, whereas the texture of the body surface is often deemed less significant—unless it is in some way elaborate and unique.

While taxonomic trends have ebbed and flowed, and species have sometimes been lumped together and sometimes split, what has stayed constant is the fantastic diversity of species. Jellyfish run the gamut of the imagination: they may be large or small; cerulean blue, lemon yellow, shades of red, or purple and silver; lethal or innocuous; disk-shaped, box shaped, long stingy stringy thingies, or shaped like the Batman logo. However, it is clear from modern analytical methods, including cladistics and molecular analysis, that the true diversity of jellyfish has been significantly underestimated. Numerous surprising discoveries of new species—even in populated coastal areas—suggests that we have only just begun to scratch the surface of understanding these unusual beasts. Once discovered and taxonomically classified, we can begin to explore their individual nature and habits.

# PUTTING TOGETHER TAXONOMY AND PHYLOGENY

Today, even with the advent of molecular analytical tools, there is still an inherent mismatch between the way we instinctively seek to identify species and the way we scientifically seek to understand their evolutionary relationships. When we look at species in a field guide, or identify them using a dichotomous key, or classify them, we base our decisions on how one species differs from another. But when we seek to compare their DNA or understand their evolutionary relationships, our decisions are based on their shared characters. These different approaches can lead to confusion.

EACH SPECIES in its earliest stages begins from within another species. As species evolve, sometimes changes occur that are so major, that they no longer resemble or function like their ancestors. These are what we recognize today as higher taxonomic groups. Darwin wrote in *The Origin of Species*, "There is grandeur in this view of life." If lineages didn't modify with descent, then we would all just look like bacteria.

## Two Systems

Systematists, or scientists who study species and their relationships, use two primary tools to express these concepts. The taxonomic system uses hierarchical lists to keep track of species names and identities, while the phylogenetic system uses family trees to understand species relationships. These two systems are not in competition with each other; they tell us different things and are complimentary. Charles Darwin described evolution as "descent with modification." We observe the descent part in the similarities that come from shared genetics, while we observe the modification part in the differences that come from mutations and other changes.

Most workers aim to have a "natural" hierarchy of biological names that attempts to reflect evolutionary origins and relationships. Unfortunately, we still know so little about the evolutionary histories of so many groups of organisms that the nomenclatural hierarchy remains, by necessity, "artificial." For example, if an order of organisms is found to have evolved from within another order or family, then the classification is adjusted to elevate the ancestor to a higher taxonomic level or reduce the descendant to a lower level.

An often-used example is reptiles and birds. There was a time far back in the history of animal life when reptiles were considered a crown group—that is to say, no other "fundamentally different group" had yet evolved from them. They were a grade of organisms that all crawled on their bellies, which had evolved this way because of their shared ancestry. Birds eventually evolved from within this group, but are distinctly different from their ancestral grade. Some would say, therefore, that "reptiles" must include birds because they are natural descendants.

Although matings and generations occur at a fairly steady rate, the way changes are reflected—or evolution—is not so steady. Sometimes it is quite static for long periods.

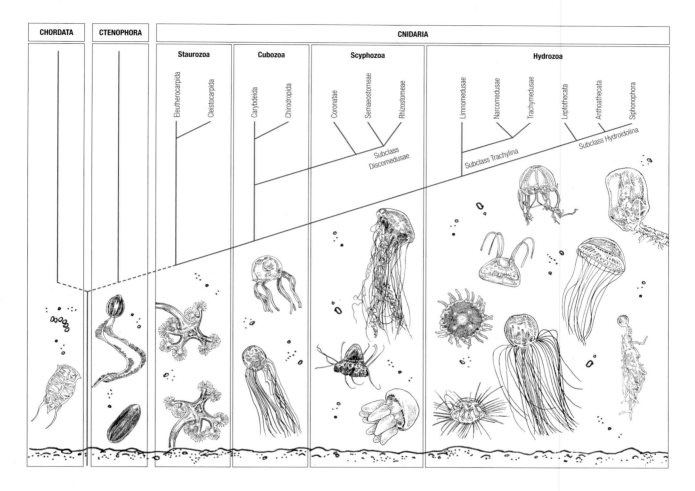

Occasionally, big glorious changes occur, which fundamentally change the whole nature of the descendants relative to the ancestors. For example, backbones, multicellularity and a central nervous system are other examples of not just descent, but major modification.

The jellyfish have a similar problem to that of the reptiles: the siphonophores are so different from the "medusozoan" jellies in their biology and ecology that they have long had their own specialist workers with their own classification system, and the group Siphonophora has been variously interpreted at the class and order levels. However, it is now thought that this group evolved from within the order Anthoathecata.

**Above** What we commonly call "jellyfish" covers an incredible array of species from three distantly related phyla and 27 orders. How we perceive them often depends on whether we focus on their similarities or differences. Studies of their DNA allows us to understand their evolutionary relationships. However, incredible leaps in form and function—like that of the siphonophores from their hydroid ancestors—challenges the limitations of our classification.

Can a class or an order have its origins from within another class or order? This would seem to deny our linear view of evolution. But the Siphonophora is such an extraordinarily bizarre group of organisms that it seems to defy normal explanation! The growing body of research into genetics and the mechanisms of change and inheritance may lead us to reevaluate the way we understand evolutionary process.

# BLACK SEA NETTLE

WITH ALL THE SCIENTIFIC EMPHASIS on DNA studies, climate change, and medical research today, one might think that we should know most of the species in nature and most of their biology, and that there really is not much left to discover. Nothing could be further from the truth. It is estimated that millions of species still await discovery, many of which are destined to become extinct before even being classified. Most of these are small and inconspicuous or are hidden away in such places as remote rain-forest thickets.

## Enormous yet Unseen

It came as a shock, therefore, when *Chrysaora achlyos* (Greek for "dark and mysterious sun-god") was classified in 1997. It is big: a meter in diameter and eight meters long (about three feet wide and 26 feet long). It is conspicuous: deep, dark purple, the color of burgundy wine. And it is very definitely not hidden away in some remote place: the species drifted right past the marine biologists and marine institutes in southern California and washed up on Los Angeles Harbor Beach.

Not only was this jellyfish new to science, a whole new structure was discovered with it. *Chrysaora achlyos* has evolved large internal anchoring structures that allow it to swim headfirst into strong oceanic currents without having its oral arms sheared off.

Through the years, this species had been seen by countless beachgoers, and beautiful photographs of it were published. So how did something so magnificently huge and colorful escape notice all this time? The answer is simple: the expertise to recognize it as new to science was lacking. One must wonder how many similarly conspicuous species have been overlooked around the world.

**Scientific name** *Chrysaora achlyos*

**Phylogeny** PHYLUM Cnidaria / CLASS Scyphozoa / ORDER Semaeostomeae

**Notable anatomy** dome-shaped body with 24 tentacles distributed in eight groups of three each, with four long, intertwined, pleated oral arms

**Position in water column** epipelagic, in neritic zone

**Size** bell to 1 m (3 ft. 3 in.) in diameter; oral arms to about 8 m (26 ft.) long

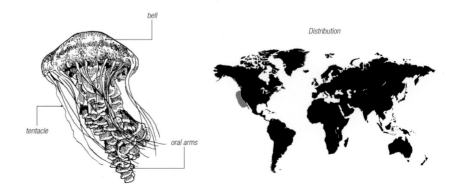

bell

tentacle

oral arms

Distribution

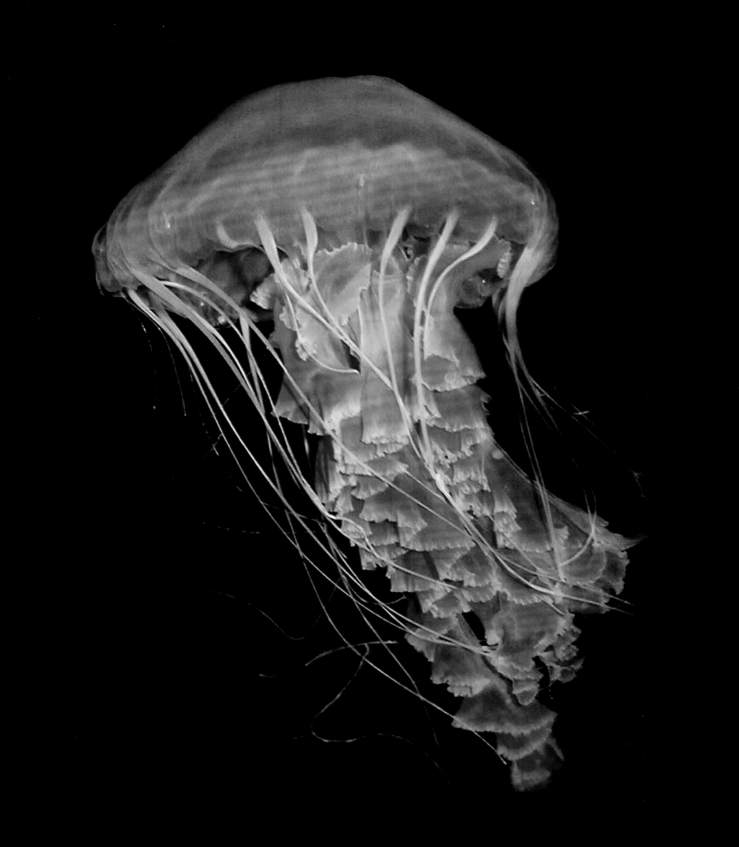

# BAZINGA!

WHILE EVERY NEW SPECIES is a wondrous thing, some stand out as a bit more remarkable than others. One such notable creature is *Bazinga rieki*, the smallest known blubber jelly, only just discovered in 2013. *Bazinga rieki* is so different from all other species that it was immediately recognized as a new species, new genus, new family, and new suborder.

## A Different Kind of Blubber

Like other blubber jellies, *Bazinga* lacks true tentacles and feeds with hundreds of tiny mouthlets rather than one main central mouth. However, it differs from all other blubbers in several fundamental characters, including its curtain-like oral arms, hooded sensory organs, unbranched canal system, and large stomach. It looks the same as others superficially, but the resemblance ends at the surface.

*Bazinga* pulsates more than 200 times a minute but does not cover much distance, which suggests it uses its movements for orienting in the water column rather than for dispersal. Its tissues are heavily colonized by symbiotic algae, which supply its energy needs. For now, we are limited to informed speculation about its habits. Because it has been discovered so recently, very little research has been performed on the extraordinary little *Bazinga*.

As unusual as it is to find a species so completely different from every other, even more surprising is the site of *Bazinga rieki*'s discovery. It was found near Sydney, Australia, in an area popular for swimming and diving, and the species appears to be common in the region. Evidently, because of its small size, *Bazinga* had been mistaken previously for juveniles of other species.

**Scientific name** *Bazinga rieki*

**Phylogeny** PHYLUM Cnidaria / CLASS Scyphozoa / ORDER Rhizostomeae

**Notable anatomy** small body, with a granulated upper surface; the radial canals are unbranched and parallel

**Position in water column** epipelagic, in neritic zone

**Size** bell about 2.5 cm (1 in.) in diameter

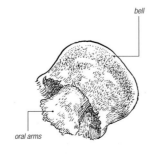

bell

oral arms

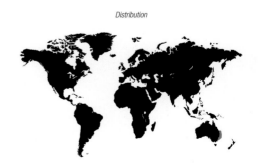

Distribution

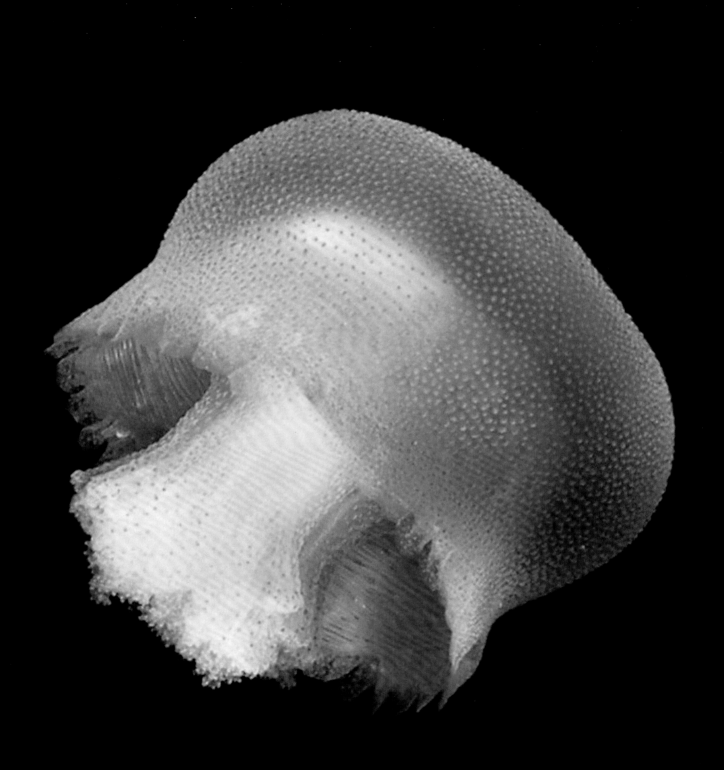

# BONAIRE BANDED BOX JELLY

ALTHOUGH DISCOVERY of new species is more commonplace than most people may imagine, it is not often that a conspicuous species is revealed to be new to science. When this occurs, it is naturally a great thrill to the discoverers and of great fascination to the general public.

### Teacher-Scientist
One such fabulous find was a large and apparently dangerous box jellyfish in the waters surrounding the Caribbean island of Bonaire. It attracted the attention of a Florida high school teacher, who became convinced that it was likely to be a new species.

The teacher compiled a list of 50 sightings of this unknown box jellyfish from various localities throughout the Caribbean. He also brought it to the attention of scientists at the Smithsonian Institution in Washington, DC, who became just as intrigued with it.

Very little is currently known about the ecology of this new cubozoan, except that it is generally observed foraging during the day. It is only known to have stung three people, one of whom required hospitalization.

### Name That Species
In searching for a name for this new species, the decision was made to engage the public in the naming process. People were invited to submit ideas online for names as part of the 2009 Year of Science commemorations.

The species was already slotted into the genus *Tamoya*, based on its genetic affinity to other species in this group. The winning submission for the species name was *ohboya*, based on what its discoverers presumably exclaimed when finding it. *Tamoya ohboya* is not only a great name, but this species' story is also a great example of citizen science.

**Scientific name** *Tamoya ohboya*

**Phylogeny** PHYLUM Cnidaria / CLASS Cubozoa / ORDER Carybdeida

**Notable anatomy** tall, box-shaped body with four flattened tapeworm-like tentacles; the tentacles are striped with dark brown to reddish orange

**Position in water column** epipelagic, in neritic zone

**Size** bell to 15 cm (6 in.) in height

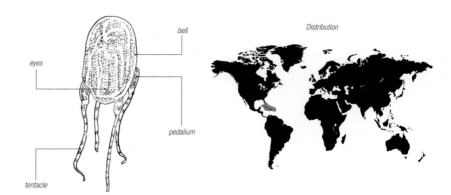

eyes

bell

pedalium

tentacle

Distribution

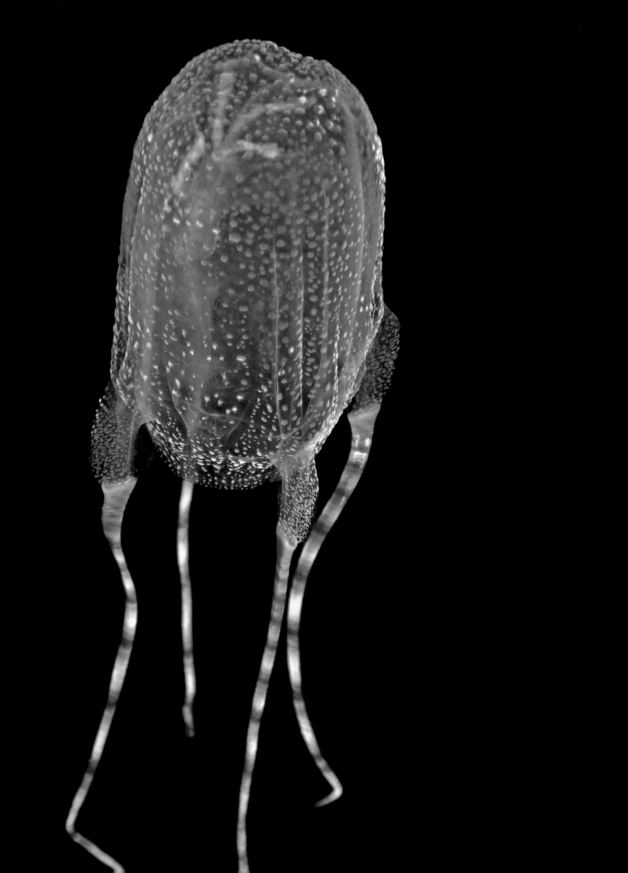

# PURPLE ROPY JELLY

A SPECTACULAR SPECIES OF JELLYFISH washed up on a popular swimming beach in Australia during the summer of 2014. The creature was a brilliant purple—almost neon bright—and had four long, ropy oral arms. Its body was hemispherical and almost a foot in breadth.

## Identity Unknown

The mystery jellyfish was immediately recognized as a species in the genus *Thysanostoma*, but one that had not previously been reported in Australian waters. The story of its finding made headlines around the world, and the publicity led to many additional reports of the species up and down Australia's eastern seaboard. It became clear that the species was not so rare, just rarely reported. Clarifying its identity, however, was less straightforward.

The possibilities in the genus *Thysanostoma* were narrowed down to two species, one from Malaysia and the other from the Red Sea. Or it might be new to science. Both of the possible species were named and classified in the 1800s. With their reference specimens long gone and their written descriptions lacking focus on characters used today, it has been difficult to say confidently whether the Australian form is one of the known species.

As is so often the case with new discoveries, almost nothing is known about this beautiful species. One of the few things that is known is why it took so long to be reported: many of those submitting photos said that it caught their attention because it was so beautiful, but they never imagined that their find was scientifically important.

**Scientific name** *Thysanostoma* spp.

**Phylogeny** PHYLUM Cnidaria / CLASS Scyphozoa / ORDER Rhizostomeae

**Notable anatomy** hemispherical body with four long, ropy oral arms; bright purple

**Position in water column** epipelagic, in neritic zone

**Size** bell to about 25 cm (10 in.) in diameter

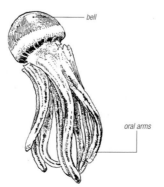

bell

oral arms

Distribution

# THREE-LEGGED BOX JELLY

The three-legged box jelly (*Tripedalia cystophora*) has a history full of poignant loss and daring world travel. It was originally encountered in Kingston Harbor, Jamaica, where its intrepid discoverer sadly succumbed to yellow fever while studying the species. Later, the mangrove swamp where *Tripedalia* lived was ripped up, filled in, and built over with hotels in the name of progress. The species is now thought to be extinct in Jamaica.

### World Traveler

But then *Tripedalia* began popping up all over the world. It is easy to recognize by the three tentacles—each with a separate unbranched pedalium (tentacle base)—on each of the four corners of its tiny box-shaped body. Presumably it spread in ballast water as ships filled their hulls at one port and discharged the water at another.

*Tripedalia* is the only known box jellyfish that can truly be considered harmless to humans. Its lack of virulence is not a matter of size: others in its diminutive range can be mighty dangerous, such as *Carukia barnesi*, whose sting causes Irukandji syndrome (page 154); and *Copula sivickisi*, the sting of which can cause a herpes-like reaction (page 88). *Tripedalia* simply does not need to be dangerous to vertebrates, because it preys primarily on minute copepods, which it hunts between the light shafts in mangrove thickets. *Tripedalia* uses its well-developed eyes to navigate among mangrove roots, around rocks and corals, and away from predators.

Most of what we know today about box jellyfish vision and mating comes from studies on *Tripedalia*—it is a veritable lab rat of the jellyfish world.

**Scientific name** *Tripedalia cystophora*

**Phylogeny** PHYLUM Cnidaria / CLASS Cubozoa / ORDER Carybdeida

**Notable anatomy** small box-shaped body with three separate tentacles on each corner

**Position in water column** surface waters in mangrove swamps

**Size** bell less than 1 cm (0.4 in.) in diameter

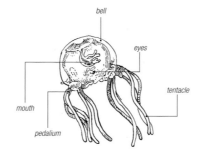

bell

eyes

tentacle

mouth

pedalium

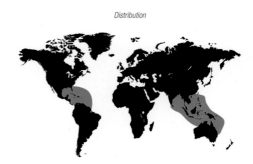

Distribution

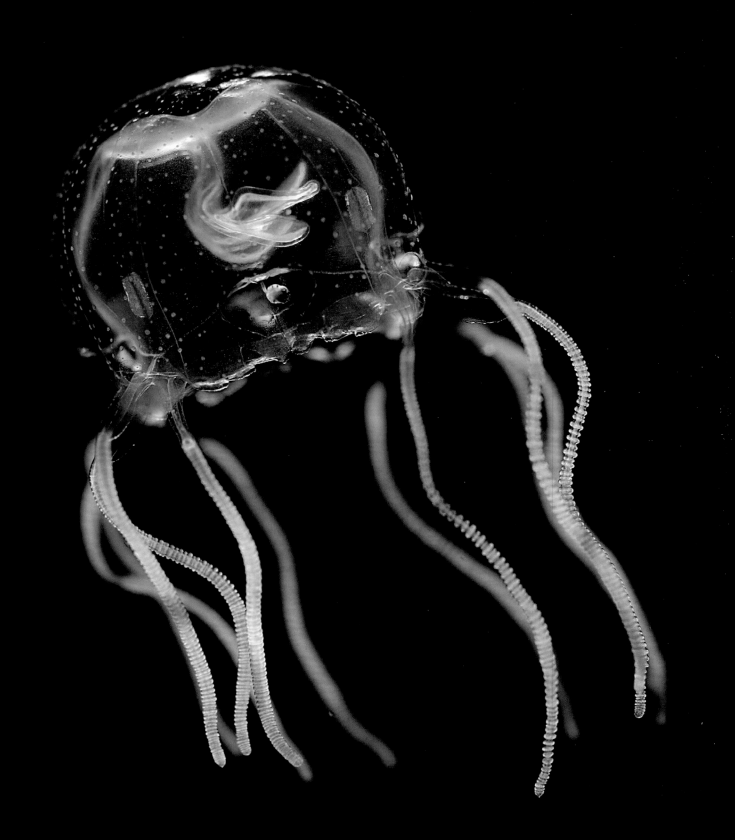

# A PELAGIC STAUROMEDUSA

STAUROZOANS ARE GENERALLY considered an entirely benthic form of medusa—that is, they dwell on the seafloor attached to a substrate throughout their life cycle and lack a pelagic stage. But there is one species that challenges this convention. *Tesserantha connectens* is thought to be a pelagic staurozoan.

## Found and then Lost

*Tesserantha* was classified in the late 1800s from four specimens collected in deep-sea trawls near Chile aboard the great oceanographic expedition of the HMS *Challenger*. The species has never been found again.

The specimens were said to be less than a centimeter (two-fifths of an inch) in height, and a little over half that in width, and characterized by a tall, pointy projection at the apex of the bell. The bell's rim bears 16 simple tentacles, rather than the more typical staurozoan form of eight arms, each bearing a tuft of short, knobbed tentacles. It is unclear why this species was classified as a staurozoan, but its description was thorough enough that it would no doubt be recognized if found again.

The most surprising thing about *Tesserantha* is that scientists of the twentieth century believed it to be apocryphal. They made the odd assumption that one of the expedition members invented the species, rather than considering the more reasonable probability that it may exist, but its discoverer simply erred in its assignment as a stauromedusa. Doubting its very existence, they discarded it from the registers of species. On the one hand, ignoring it simplifies things by eliminating the mystery of why this purported pelagic staurozoan has never been found again. But on the other hand, it makes it less likely that the species will be recognized for its significance if found again, and the mystery of its identity truly solved.

**Scientific name** *Tesserantha connectens*

**Phylogeny** PHYLUM Cnidaria / CLASS Staurozoa / ORDER Tesseranthida

**Notable anatomy** small round body with a hollow apical projection; 16 solid tentacles

**Position in water column** abyssopelagic, found at 4,200 m (about 14,000 ft.)

**Size** bell 9 mm (0.35 in.) in height, 6 mm (0.25 in.) in diameter

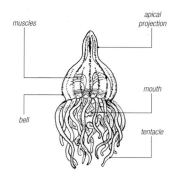

muscles

apical projection

mouth

bell

tentacle

Distribution

# ENIGMATIC FOSSIL "MEDUSAE"

LANDLOCKED SANDSTONE QUARRIES are hardly the sort of place one might expect to find jellyfish, but splendid discoveries have been unearthed in these sites in many places around the world. Some clearly appear to be jellyfish, while others are less certain and open to reasonable debate.

### Stranded in Wisconsin
For example, credible medusa fossils have been found in a central Wisconsin quarry used for the mining of paving stones and kitchen counter slabs (page 99). These are not just a few prize finds—these are vast expanses of rippled sand flats dotted with stony lumps, each lump formed by jellyfish that were washed ashore and entombed together on an ancient sandy beach. What's more, seven consecutive bedding planes of these fossilized jellyfish blooms are stacked like pages in a history book, dating back some 510 million years.

### The Ranford Fossils
Of a more ambiguous origin are spectacular star-shaped fossils from Western Australia. The Ranford Formation where these were found is better known for zebra rock—a dazzling deposit of distinctive reddish-brown and white-banded siltstone. Within this deposit are numerous small disk-shaped impressions comprised of radiating, stiff, straight-sided shards. Some experts have argued that these medusa-like fossils are not of biogenic origin, that is to say, not biologically generated. Others have argued that they are, based on their uniform size and structural pattern. It has been further argued that these may have been similar to the keratinized disks of the Blue Button *Porpita* (page 78). These fossils derive from a glaciated period in the Precambrian called the Cryogenian; if biogenic, they would reset the date of the origin of animal life.

**Scientific name** unclassified

**Phylogeny** PHYLUM Cnidaria / CLASS unknown

**Notable anatomy** small disk with radiating, shard-shaped segments

**Position in water column** unknown

**Size** to about 2.5 cm (1 in.) in diameter

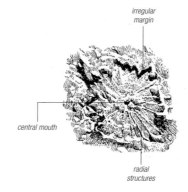

irregular margin

central mouth

radial structures

Distribution

# WHEEL SALPS

NLIKE MOST SALPS where the aggregate members are stuck together in species-specific chain-like arrangements of end-to-end, herringbone, and so forth, cyclosalps are arranged around a common point and stuck together by stalks like the spokes of a wheel. What's more, the wheels may be joined together side by side into long chains, like a string of very funny looking beads.

## Unusual Features

Besides their wheel form, cyclosalps bear conspicuous structural differences from "normal" salps and are easy to tell apart with even casual observation. In the regular salps, the gut is globular at one end of the body in both the solitary and aggregate forms. But in cyclosalps, the gut is linear through the body in the solitary stage and U-shaped in one end in the aggregate stage.

Cyclosalps are similar to normal salps in their general ecology. Both are voracious grazers on phytoplankton, which they filter through a mucous net that they produce continuously. The mucus is fed into the mouth like a conveyor belt, bringing captured food particles.

Similarly, cyclosalps are like regular salps in their biology. The solitary zooid asexually buds a chain of young salps. These are all female, which become male as they age. Young female chains are fertilized by older male chains. Successful mating produces an embryo inside the fertilized female.

Bioluminescence is well known in the pyrosomes and appendicularians (distant kin of cyclosalps), and cyclosalps have been sometimes reported as bioluminescent. However, this appears to be erroneous and they are thought to not be capable of producing their own light.

**Scientific name** *Cyclosalpa* spp.

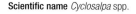

**Phylogeny** PHYLUM Chordata / SUBPHYLUM Tunicata / CLASS Thaliacea / ORDER Salpida

**Notable anatomy** barrel-shaped zooids attached in a wheel shape in the aggregate form

**Position in water column** epipelagic, in open ocean and neritic zone

**Size** body to about 15 cm (6 in.) long, usually much less

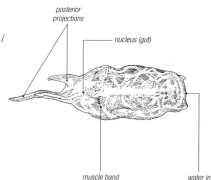

posterior projections

nucleus (gut)

muscle band

water in

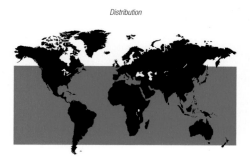

Distribution

# MOON JELLYFISH

UNTIL RECENTLY, the genus *Aurelia* was said to consist of a single species—the Moon Jellyfish (*Aurelia aurita*)—that was cosmopolitan in its distribution and found in nearly all coastal habitats. However, it is becoming increasingly clear that there are many local varieties of *Aurelia*, and the interpretation that it is a single global species is now being questioned. It appears that there may, in fact, be dozens to hundreds of species.

### Distinctive Appearance

*Aurelia* is one of the most recognizable of all jellyfish: the body is fairly flat, usually ghostly whitish in color, with typically four large horseshoe-shaped rings, conspicuously visible, near the center. The rings are the reproductive organs, and the area inside of each ring is one of the four stomachs. The margin of the body is scalloped, and fringed with hundreds of short fine tentacles used for catching tiny planktonic organisms for food.

Despite having so many tentacles, *Aurelia* is generally harmless to humans, or nearly so. Its sting is barely felt by most people; if perceived, it feels like the application of a warm blanket. The sting has even been used therapeutically to treat arthritis, its use presumably based on the belief that the microscopic stings stimulate blood flow.

Humans have a love-hate relationship with *Aurelia*. It is loved because it is so beautiful, and indeed it is among the most popular exhibition animals at public aquariums around the world. But it is hated because its blooms are a source of endless havoc, causing emergency shutdowns of power plants and mass fish kills at salmon farms.

**Scientific name** *Aurelia* spp.

**Phylogeny** PHYLUM Cnidaria / CLASS Scyphozoa / ORDER Semaeostomeae

**Notable anatomy** flattened body with scalloped edge and hundreds of short, fine tentacles; bell with four conspicuous internal rings

**Position in water column** shallow, coastal; generally found in bays and harbors

**Size** mature bell 30–50 cm (12–20 in.) in diameter

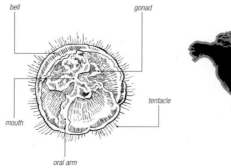

bell — gonad — tentacle — mouth — oral arm

Distribution

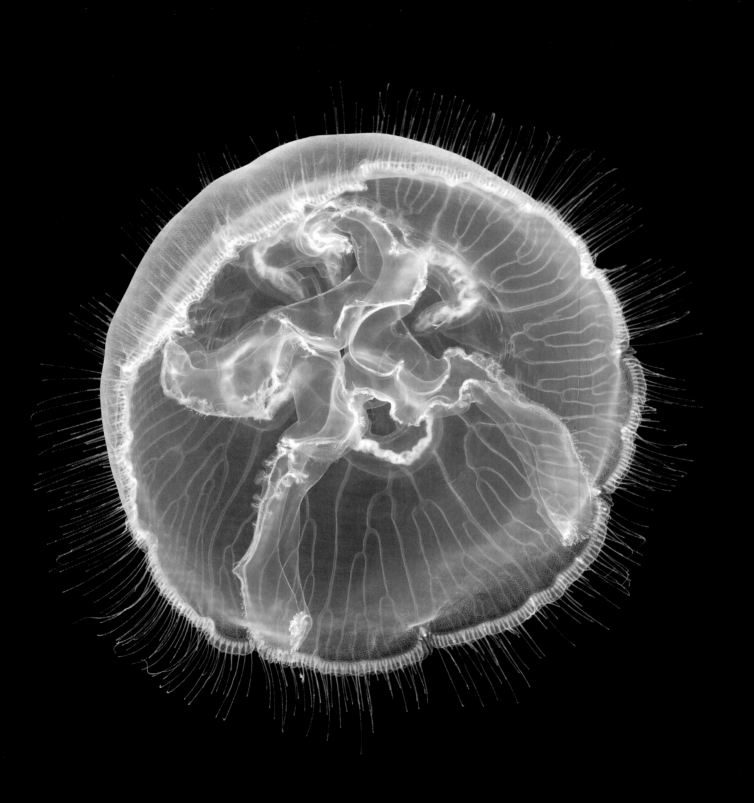

# VENUS'S GIRDLES

WHILE MOST CTENOPHORES can justifiably be described as strange—or even, not inappropriately, as alien—there can be no doubt that the strangest, most alien of all are *Cestum* and its little cousin *Velamen*. These two species, known as Venus's girdles, are so distinct from other ctenophores that they have been placed in their own order, Cestida.

The Venus's girdles are flattened in the extreme, with the mouth midway along one of the long edges, and the majority of the body drawn out into two long lobes, resembling a belt. Four of the comb rows are rudimentary; the other four are arranged along the edges of the lobes on the aboral side (the long edge opposite the side with the mouth).

*Cestum* and *Velamen* can be distinguished from one another easily when mature: in *Cestum* the gonads appear as a continuous band along the aboral edge; in *Velamen* the gonads are in a broken line.

## Different Strokes

It might seem logical for a belt-shaped organism to move through the water like an eel, with one end leading, but the Venus's girdles move this way only as an escape response. When swimming normally they move mouth-first. Two rows of fine tentacles are attached in grooves along the entire oral edge of the body; these are used for capturing small food particles such as copepods.

## Rarely Seen

The Venus's girdles are found in all oceanic waters, and occasionally are brought into coastal waters by onshore currents. They are incredibly delicate and shatter when their body strikes a collecting net. Therefore, they are typically observed only in situ by a lucky few.

**Scientific name** *Cestum veneris* and *Velamen parallelum*

**Phylogeny** PHYLUM Ctenophora / CLASS Tentaculata / ORDER Cestida

**Notable anatomy** long, ribbonlike body

**Position in water column** epipelagic, mesopelagic, bathypelagic

**Size** body length to 1.5 m (60 in.) in *Cestum*, to 20 cm (8 in.) in *Velamen*

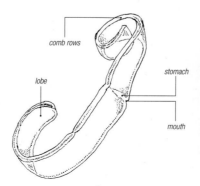

comb rows

lobe

stomach

mouth

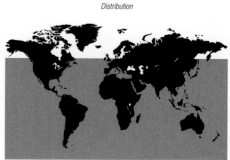

*Distribution*

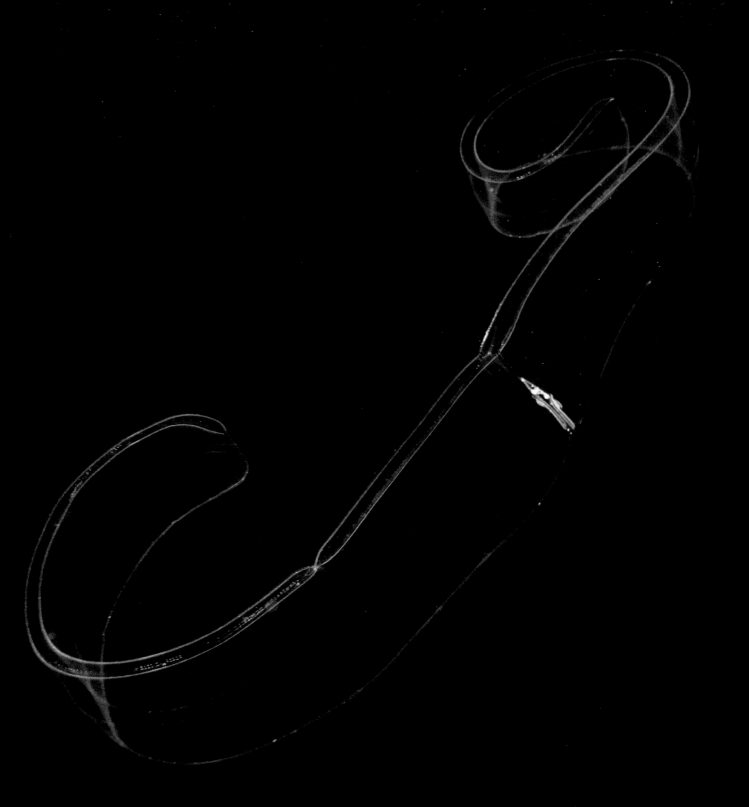

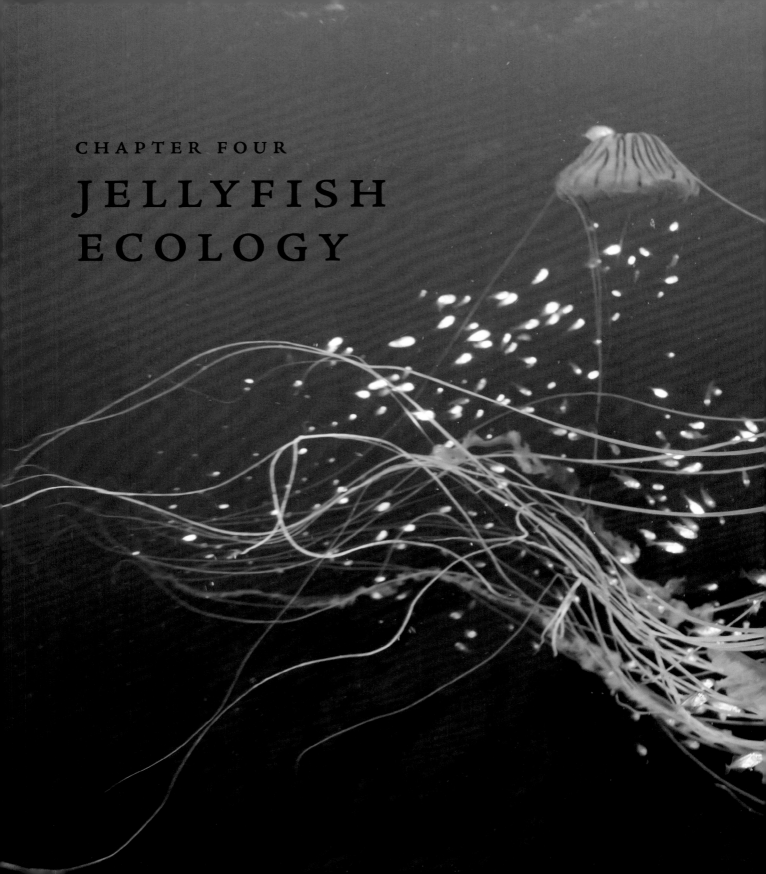

CHAPTER FOUR

# JELLYFISH
# ECOLOGY

# JELLYFISH ECOLOGY: AN INTRODUCTION

Ecology is the study of how organisms interact with one another and with their environment. Jellyfish are simple animals that have been around an incredibly long time—some 600 million years or more. During this time, while other species have come and gone, evolved lungs or legs or wings, learned to walk or fly, jellyfish have not changed. They have not needed to. Quite simply, what they do (and how they do it) works. Their ecology and biology, their predator-prey relationships, and the environmental cues that drive their behaviors have all been finely honed over hundreds of millions of generations.

### The Jellyfish Success Story

We often think of jellyfish only as pests. They sting us when we go to the beach. They inundate fishing nets, and clog the cooling intake pipes for nuclear power facilities and desalination plants. They compete for food with fish, seabirds, and marine mammals, and ultimately with us, because we harvest from the sea, too. Sometimes they drive ecosystem shifts that result in less-desirable states. But we take a myopic view if we look only at the problems jellyfish cause. There is much more to these beautiful and mysterious organisms, which we are only beginning to know, including the qualities that make them the successful pests they are. Indeed, they are admirable for their adaptability, their tenacity, and their persistence.

The organisms that we place in the group "jellyfish" are of incredible diversity and collectively occupy a unique position in the ocean: jellyfish are predator, prey, competitor, or companion of just about every marine creature. Some jellyfish

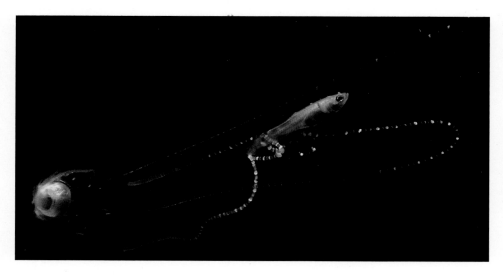

**Left** Jellyfish interact with other species in their environment in many different ways, most conspicuously as predator and prey. The life-threatening Irukandji *Carukia barnesi* has a voracious appetite for fish, rolling one up inside the stomach while fishing for more.

**Above right** Many larger jellyfish act as mobile homes for other species. Juvenile trevalla (*Caranx* spp.) are frequent associates of jellies, using them as both a food source and protection. Crabs, barnacles, and even brittle stars have a similar relationship.

eat only phytoplankton; others eat only zooplankton; still others eat only each other. Many will eat anything. Together they can take over an ecosystem. In this chapter we examine some of the ecological relationships, behaviors, and specializations that allow jellyfish to survive, to flourish, and to dominate, from their seasonality and bloom dynamics to the sophisticated migrations they perform on a daily basis to the ways they hunt their food or avoid becoming food. Drifter, sailor, pest, or hitchhiker host, jellyfish play many roles and are many things to many species.

## The Importance of Jellyfish Studies

For most of modern scientific history, marine biologists have ignored jellyfish, considering them to be lowly and irrelevant. Scientists on expedition ships have commonly thrown overboard the jellies brought up in a sample to make way for the fish they want to study. Even when faced with a sample that is 90 percent jellies, most of the time scientists have simply waded through them rather than thinking about the impact that so many mouths and so much biomass might be having on the oceanic ecosystem. It has been only since 1995 that scientists have realized that jellyfish can be both a visible indicator that the ocean is out of balance and a driver of further decline.

We stand now at the beginning of a new era. The oceans are changing fast: temperature, pollution, and coastal use are increasing, while oxygen, pH, and biodiversity are going down. And jellyfish are revealing themselves to be major players in the dynamics of the marine environment. They are the inheritors of these rapidly changing ecosystems; after the species we favor have been fished out or killed off, jellyfish remain. In some situations, the jellies themselves play the angel of death. In others, they are the last man standing. Understanding the ecological principles that govern their lives and control their populations may help us to better deal with the problems they cause and to develop mitigation strategies.

For sound scientific and industrial reasons, we should pay attention to jellies—but another reason is simple curiosity. Jellyfish are amazing: beautiful but deadly, simple yet enduring. We know little about them and the way they interact with one another, with other species, and with their environment, but they surely harbor fascinating secrets. Scientists who study them know we can expect only more eyebrow-raising discoveries, more stories that border on the unbelievable.

Although jellyfish seem simple at first glance, a closer look reveals the complexities of their ecological relationships and of the biological processes that enable them to thrive in what must certainly be, for them, a dangerous world.

# PREDATION AND COMPETITION

Jellyfish have been in the news in recent years, as their blooms—rapid increases in numbers in a particular area—have caused operational challenges and financial losses for many industries. One of the biggest problems, however, is less discussed: the effect that jellyfish blooms have on other organisms in the ocean. Often jellyfish blooms are the result of ecosystem decline, and, moreover, jellyfish can—and sometimes do—cripple ecosystems. We discuss the causes of these blooms in depth in chapter five. Here we look at one of the main mechanisms by which jellyfish can actually drive ecosystem decline.

### How Jellyfish Take Control

Jellyfish act as both predators and competitors of numerous other species—including other jellies and many kinds of fish. They even compete with animals as large as sharks and whales for food. Jellyfish eat the eggs and larvae of other marine species, as well as the food that those larvae and adults would eat. For instance, they eat the larvae of the krill upon which so many animals (including baleen whales) depend, as well as the phytoplankton that the krill eat. Jelly blooms apply huge predation pressure near the base of the food web, and these impacts ripple up the food chain to affect even mammals, birds, and large fish that seemingly have nothing to do with jellyfish. The double impact of predation and competition enables jellyfish to take control of an ecosystem and restabilize it, creating a "new normal." This has occurred many times in different habitats around the world, suggesting a common pattern.

Jellyfish conquer an ecosystem by different pathways, which become available with the occurrence of certain conditions or combinations of conditions, such as overfishing, excess nutrients, warming water, and so on. And once an ecosystem has flipped to dominance by jellies, this simplified condition seems to be remarkably resistant to changing back to one of greater complexity.

### The Jelly Web

Scientists have recently described a secondary food chain—the so-called jelly web—that operates parallel to the more conventional food chain. Jellyfish can thrive with very low expenditure of energy. Whether they eat each other or microbes or dissolved nutrients or sludge on the bottom or nothing at all, they still thrive. They do not need a food chain containing more complex animals, such as crustaceans, mollusks, and fish. In fact, jellyfish evolved and flourished for millions of years before these other species even appeared. Jellyfish occupy the niche of top predator in this low-energy food chain. The conventional oceanic food chain supports organisms with higher energy needs, such as fish, whales, seabirds, sharks, and people. These two food chains are not entirely exclusive of one another, however; jellyfish are happy to eat from either menu.

One of the reasons jellyfish present such a serious threat to oceanic ecosystems has to do with energy conversion. Jellyfish convert energy the wrong way around—or at least not the usual way. Normally, higher-energy organisms eat lower-energy food. For example, a pound of beef packs a higher energy content than a pound of grass, beef cattle's common food source. Similarly, a pound of tuna packs more energy than a pound of the tuna's crustacean prey, and a

PARALLEL FOOD CHAINS

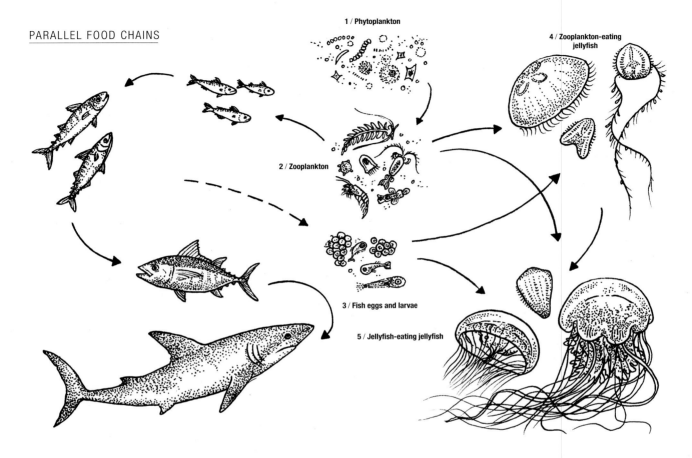

**1 / Phytoplankton**

**2 / Zooplankton**

**3 / Fish eggs and larvae**

**4 / Zooplankton-eating jellyfish**

**5 / Jellyfish-eating jellyfish**

pound of crustaceans packs more energy than the prey they consume. In general, the higher up a species is on the food chain, the higher its energy value should be. Jellyfish, however, subvert this system. Jellyfish eat fish eggs and larvae—species higher up on the food chain than they are—and convert this energy in their own bodies to a lower-value, less usable, and lower-quality food source. In jellyfish blooms, this process shifts an ecosystem from a higher-energy, fish-dominated bionetwork to one that provides less energy value to other species, including humans.

Among the primary competitors of jellyfish are the "small pelagics," or forage fish, including such schooling species as sardines, anchovies, pilchards, and menhaden. These small fish are food for just about everything bigger than they are. Whales, dolphins, seabirds, and sharks and other large fish

**Above** The conventional food chain (left) and jelly web (right) can exist independently. **1.** Phytoplankton grow from nutrients and sunlight. **2.** Zooplankton eat phytoplankton. **3.** Fish eggs and larvae become part of the plankton. **4.** Many jellyfish eat plankton including fish eggs and larvae and compete with them for food. **5.** Some jellyfish also eat other jellyfish.

all rely on them as a primary part of their diet, or their diet's diet. Several recent studies have revealed a seesaw effect between jellyfish and small pelagics in some ecosystems. When the small pelagics have been fished down or otherwise reduced, the jellyfish increase; as they consume eggs and larvae, their further impact on the pelagic population jeopardizes the sustainability of species that feed on the fish. Humans rely on these fish too, not only as a pizza topping and a canned food, but also for aquaculture feed, agricultural fertilizers, bait, and fish oil.

# JELLYFISH AS PREY

Many oceanic species eat jellyfish, and some specialize in them. Sea turtles are well-known predators of jellyfish. This relationship is one of the reasons plastic grocery bags are so harmful in the ocean: turtles often cannot tell the difference between the floating plastic and their prey. Blue Swimmer Crabs (*Portunus pelagicus*) of the Indian and Pacific Oceans swim up into the water column, grab a jelly, and then sink back down again and begin shredding the gelatinous tissues apart, shoving clawful after clawful into their quivering mandibles. Fish nibble at jellies. Dolphins play Frisbee with medusae, in most cases leaving the toys for dead in the end. Even foxes and other land mammals, as well as birds, will eat jellies stranded along the shore when other food is scarce. In many Asian cultures, jellyfish have been considered a delicacy for thousands of years.

### Defense Against Predators

All living things develop some means of defending themselves from would-be predators. Even plants have spines, thorns, and noxious or resinous chemicals for protection. Soft-bodied organisms in particular are easy prey for other species. Jellyfish tend to be slow moving and fleshy, and they lack the hard parts that their coral brethren have developed into a protective skeleton. Moreover, they have no teeth or claws with which to fight.

One of the most obvious defensive tools jellyfish have is their stinging cells, which we examined in detail in "Stinging Cells and Sticky Cells" (pages 20–21). We now turn our attention to structural and behavioral modes of defense that jellyfish employ to keep themselves from being eaten.

Some species, such as *Csirosalpa caudata*, have gelatinous projections on the body. These appear to play a role in maintaining buoyancy by creating drag that slows the animal's sinking rate. These structures may also help in defense by putting more distance between the center of the animal's body and would-be predators.

Many types of jellyfish, such as the blubbers, have very thick, bulky bodies, which allow for the occasional nibble or chomp out of reach of vital organs. Most jellies can repair damaged or missing parts in a matter of hours to days. Jellyfish also often swarm together, creating an exclusion zone to thwart other species, a strategy of safety in numbers, similar to that of schooling fish or flocking birds.

Species of fish and seabirds are commonly observed picking up a jellyfish and then spitting it back out again. This may indicate the presence of unpalatable chemicals in the jelly's tissues. Other soft-bodied creatures, including sponges and soft corals, employ this defensive strategy; many of the chemicals are under investigation for pharmaceutical usefulness. Although most jellies have not yet been assessed for the presence of these types of chemicals, these substances are most likely to be found in particularly vulnerable body parts or species, especially those without stinging cells or other defenses.

Perhaps the most unusual defensive strategy among the jellyfish is found in at least two distantly related species of comb jellies from Japan. When disturbed, they spurt an ink similar to a tincture of iodine. This probably has a similar function to the ink-cloud defense used by cephalopods (octopuses and squids), in which the ink creates both a diversionary focal point and a smoke screen to allow escape.

## Bioluminescence

Another important jellyfish defensive mechanism is bioluminescence, light created by the organisms themselves, typically emitted as a flash or a glow. Flashes may trigger a startle response in a predator, or blind or confuse it long enough for the would-be prey to swim away unnoticed. Many types of jellyfish flash, including the pyrosomes, most ctenophores, and deep-sea medusae. Glowing is more typically used by species to create a diversion; the hydromedusa *Colobonema sericeum* (page 46), for example, casts off wriggling, glowing tentacles in the way a lizard distracts a predator by jettisoning the tip of its tail.

Bioluminescence is not always used to startle prey or create a diversion. Some higher invertebrates, such as squids, can use bioluminescence to mimic the down-welling light that normally emanates from the sun, or even the moon, effectively camouflaging themselves from other species. It seems likely that at least one jellyfish species also uses bioluminescence as camouflage: the deep-sea scyphozoan *Periphylla periphylla* (page 166) displays twinkling instead of flashing lights. In the dark, this probably helps alter its appearance, so rather than one large jellyfish it seems to be many smaller organisms.

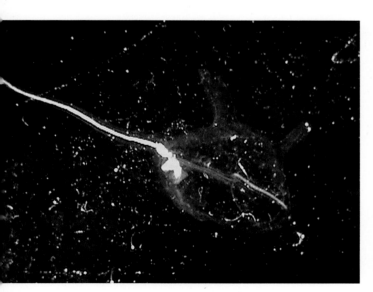

**Far left** Sea turtles, like the Hawksbill Turtle (*Eretmochelys imbricata*) seen here eating a blubber jellyfish, will take a good chomp when the opportunity arises. They can even eat the powerfully toxic box jellyfish without getting stung.

**Left** Many jellyfish have spiky or finger-like gelatinous structures that put some distance between them and their predators, and also help slow their sinking rate by creating drag. Projections in *Csirosalpa* aren't actually sharp, despite appearances.

# HIDING IN PLAIN SIGHT

In the previous section we looked at some of the structural defensive strategies that help soft-bodied aquatic organisms such as jellyfish avoid becoming easy prey. Another adaptation common among marine creatures, especially those of the open ocean, is an appearance that makes them essentially invisible through camouflage or transparency.

## Transparency

Many jellyfish are transparent. Predators and prey alike can look right through these species and not see them. Most hydromedusae, salps, and ctenophores, for example, are transparent, particularly the smaller, bite-size species. Intriguingly, the most venomous animals in the world, the box jellies and Irukandji species, are also transparent. Often, it is easier to see these animals' shadows on the sand than to see their bodies in the water. Of course by then it may be too late. *Carukia barnesi*, the Common Irukandji (page 154), is so clear that even in a jar it can be hard to see.

Public aquariums use strong side lights in jellyfish display tanks to illuminate the more transparent species. Gelatinous tissue has various properties and interacts with light in different ways. It allows light to pass right through the subject from certain angles—namely from directly above and below—but glows brilliantly when the light comes from the side. Scientists and photographers take advantage of this by illuminating subjects with side beams, but in nature, of course, the sunlight shines from above, so the animals look transparent. As an added advantage, transparency costs very little energy for the organism to make; not producing pigment is certainly metabolically cheaper than producing pigment.

**Right** Transparency is one of the most common forms of camouflage found among jellyfish. Light coming from above makes ctenophores like *Bolinopsis* virtually impossible to see, particularly for organisms with limited eyesight or reasoning ability.

**Far right** While many dread encountering the Portuguese Man-of-war (*Physalia physalis*), it is hard not to be mesmerized by its sapphire blue color, which offers camouflage for creatures that live at the interface between air and water.

## Color-based Camouflage

Nonetheless, pigment-based camouflage, including mottling and spots, is a fairly common defensive adaptation in jellyfish. Many species of blubbers, such as the Spotted Jelly (*Mastigias papua*, page 160) and the Australian Spotted Jellyfish (*Phyllorhiza punctata*, page 158), bear white spots on a brownish or greenish background. To would-be predators, or prey, this pattern probably gives the impression of a cluster of small organisms rather than one large individual. Mottled blotches are common in species that live close to the bottom, such as the upside-down jellies

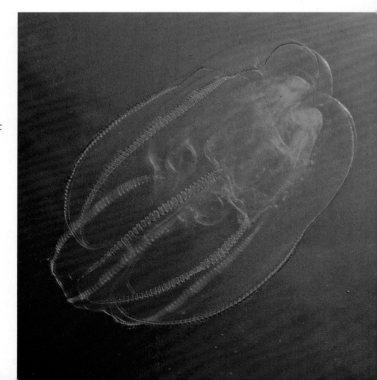

(*Cassiopea*, page 48), and especially among those that live in habitats with clear waters and visually complex backgrounds, such as coral reefs. Species that spend their time in and among reeds and algae often bear streaky pigments of different colors, which provide excellent camouflage and make them difficult even for humans to see.

Certain bold colors have also proven beneficial in some habitats. Because of the way the light spectrum travels through water, different colors penetrate to different depths. Red penetrates the least, so its dark shade looks black in even fairly shallow depths. In such deep-sea species as the Santa's Hat Jelly (*Periphylla periphylla*, page 166) and the flying saucer jellies (*Atolla* spp., page 168) the dark red coloring not only camouflages the animal, but also masks bioluminescence that may be emitted by prey in the gut. True black is an uncommon color in jellyfish, but it does occur in some species.

Bright blue is an advantageous color for species living at the air-water interface, such as the Portuguese Man-of-war (*Physalia physalis*, page 34) and the exquisitely blue *Velella* and *Porpita* (page 78). Even mollusks and crustaceans living there are this same color. This brilliant color is thought to reflect ultraviolet rays and protect the animal's tissues from sun damage. It may also reflect other wavelengths of light to help keep the animal cool. Additionally, it probably provides camouflage, making these animals essentially invisible to predators from above, such as seabirds.

In some species, color is a warning sign to would-be predators. Red and yellow are common colors that serve this purpose throughout the animal kingdom, as in North America's Common Coral Snake (*Micrurus fulvius*) and many of the poison dart frogs (family Dendrobatidae) of Central and South America. In the jellyfish group, the stinging organs of siphonophores are often red or yellow—and will cause extreme pain to anyone who ignores the message. Clusters of stinging cells on the bells of some Irukandji species are bright red or pink, possibly a similar warning sign.

Another type of color effect commonly associated with jellyfish is caused by refraction. Ctenophores, in particular, create dazzling rainbows of light with their ciliated comb rows. They are not actually making these colors, but the cilia refract available light into its component hues, which our own eyes see as the color spectrum.

# ASSOCIATIONS WITH OTHER SPECIES

Through the vast eons of time over which jellyfish have been pulsing through the oceans and seas of this Earth, they have formed associations with many other species. Not all of their interactions are as straightforward as those between predator and prey. Some relationships are symbiotic, involving two creatures living in close association with one another. These may be mutually beneficial, with each species gaining something, or may be more lopsided affairs, in some cases of which the jellyfish benefits, while in others it merely gives something away to the other species.

## Jellyfish-Algae Symbiosis

One of the most well-studied associations is that of some jellyfish and their symbiotic algae, zooxanthellae (*zoo-zan-THELL-ee*). These algae are similar to those found in corals; in both cases, the symbionts often give the host their rich coloring. Jellies that have algal symbionts, such as the upside-down jellies (*Cassiopea* species, page 48) and the Australian Spotted Jellyfish (*Phyllorhiza punctata*, page 158), are able to meet most of their nutritional needs with carbohydrates supplied by their algae. These jellies barely need to eat. As long as the sun is shining and there are plenty of nutrients in the water, the symbionts are able to photosynthesize, keeping the jellies well fed.

## Jellyfish and Other Marine Species

Another common association occurs between fish and some types of jellies, particularly the blubbers and sea nettles. These jellyfish species have lots of ruffles and crevices in which small creatures can secrete themselves away. For example, larvae of some fish, such as leatherjackets (Monacanthidae) and trevally (*Caranx georgianus*), make use of jellies much the way a turtle uses its shell. It is not uncommon to find dozens of small fish darting about beneath—or even hiding inside—larger jellies. The

Man-of-war Fish (*Nomeus gronovii*) finds protection and shelter among the long, stinging tentacles of the Portuguese Man-of-war (*Physalia physalis*, page 34), stealing food and skilfully avoiding lethal contact.

Many types of invertebrates use jellyfish as a mobile home, taking advantage of the opportunity for dispersal without having to do the swimming. Along the way, these hitchhikers benefit further by using the jellyfish for protection as well as a convenient source of nutrients, consuming food the jelly has gathered but not yet ingested. Small crabs live in the nooks and crannies between folds of tissue. Gooseneck barnacles live on the margin of the bell, hanging down between the tentacles, or on the apex of the body, sticking up like antennae. Larval sea anemones hitch rides while they develop. Fish, crustaceans, and worms live and breed inside salps. One of the most surprising associations is that between the brittle stars of the genus *Ophiocnemis* and several species of blubbers: dozens of brittle stars may be found on one jelly. Even larval lobsters make use of jellyfish: perched on the top of the bell, these developing crustaceans ride the medusa like a surfboard.

Some invertebrates go a step beyond using jellyfish merely as a protective home and larder, and use pieces of jellies or tufts of their tentacles for their own defense. For example, the

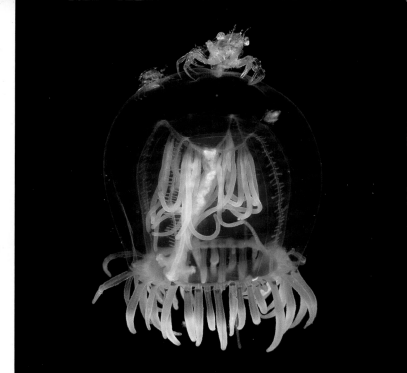

**Right** A surprising diversity of crabs, fish, sea anemones, brittle stars, and lobster larvae commonly hitch a ride on jellyfish. Some hide inside the body, while others, like this crab larva, may be found riding on top.

**Below right** Amphipods are well known for their parasitic and commensal relationships with jellyfish. Females of *Phronima sedentaria* hollow out jellies to make barrel-shaped brood chambers for their young (the red dots).

Blanket Octopus (*Tremoctopus violaceus*) lines up jellyfish tentacle fragments in single file on the suckers of its eight arms; it appears that these are used for both defense and offense. Some types of crabs will even wear a jellyfish as a protective jacket, holding the hapless medusa in place with their smaller legs.

## Parasitic Relationships

Some symbiotic relationships among jellies and other species are parasitic; in these cases the guest is destructive to the jellyfish host. Among the most common of these associations is that of jellies and hyperiid amphipods, small buglike crustaceans that burrow into the tissues of the host. Often the parasite eats the host's tissues, particularly the denser tissues of the stomach or the reproductive organs. One of the most well-studied of these involves females of the amphipod genus *Phronima*, which hollow out the insides of a salp or ctenophore, shape the outer skin into a sort of barrel, secrete a spit-like substance to harden it to the consistency of cartilage, and then lay their eggs inside. The female *Phronima* cares for her developing young as she paddles the barrel through the ocean. When old enough, the young are assured a nutritious first meal by consuming the barrel, effectively eating their mother out of house and home.

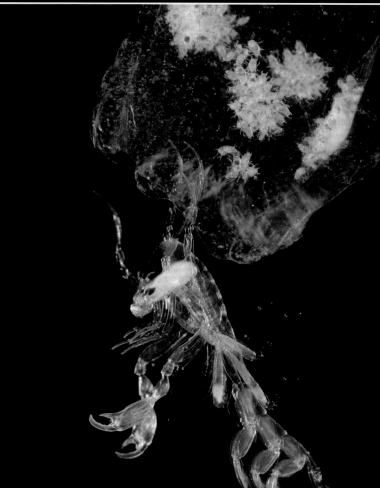

# SEASONALITY AND OTHER OCCURRENCE PATTERNS

In general, jellyfish bloom in the spring, reproduce during the summer, and die off in the fall; polyps, where present, typically endure through the winter. However, within this general pattern, different species of jellies have different occurrence patterns, just like we see with any other group of organisms. Some bloom briefly while others persist for months, and a few are periodic. Some of the more intriguing ones may live for several years or more.

## Triggers of Population Patterns

Temperature fluctuations and changes in day length are less pronounced in the tropics and subtropics; in these regions, species abundance is often strongly associated with other conditions that occur during different seasons. In tropical Australia the so-called stinger season is generally assumed to coincide with the hot, humid conditions of the wet season (summer). However, certain species, including some of those deadly to humans, occur or even reach their peak abundance during the cooler months of the dry winter season.

Besides latitudinal patterns, most jellyfish species also have their own particular seasonal abundance pattern. Summertime is the peak of abundance for hydromedusae as a whole, but most species do not live the entire summer, and some may live only a matter of days. Different species follow different calendars, so a succession of species appears throughout the warmer months. Species surveys often do not sample frequently enough to interpret these patterns, so our knowledge of these dynamics is limited and the prevalence of such patterns is often underestimated.

Salps and their kin often bloom rapidly and then die back when their population exceeds the food supply. These blooms may occur several times throughout the year, depending on conditions. For example, a good rain will often wash enough nutrients into coastal waters to trigger a phytoplankton bloom, which in turn triggers a salp bloom. Along-shore winds will often stir up the seas and trigger an upwelling, providing another pulse of nutrients. Other events such as sewage spills or aquaculture maintenance can also provide enough nutrients to trigger a localized phytoplankton response, which may be followed by salps or medusae.

Some species of jellies also have a daily pattern. In many protected localities such as bays and harbors, moon jellyfish (*Aurelia* spp., page 130) rise en masse to the surface of the

water once or twice a day, and then sink back down again. These aggregations are believed to be related to spawning; the animals are brought closer together, and the sperms and eggs are concentrated by the surface of the water, which acts as a barrier. These swarms are quite a sight to behold, with the medusae packed in cheek by jowl, in many cases as far as the eye can see. Similarly, many hydrozoans spawn at either sunrise or sunset each day; some aggregate at the surface, although others do not.

A fascinating occurrence pattern is that of species in the cubozoan genus *Alatina*. The stings from these species often cause the life-threatening medical condition known as Irukandji syndrome in humans (page 154). About a dozen species of *Alatina* are known, living in reefs and archipelagos throughout the tropics and subtropics of the world, from the Great Barrier Reef to Grand Cayman. These jellyfish are generally rare most of the time. However, around 10 days after the full moon each month, *Alatina* clusters in what are believed to be spawning aggregations. In Hawaii and the Caribbean these swarms occur during the daytime; in Australian waters they occur at night.

## Long-term Population Cycles

While daily, monthly, and seasonal periodicity in jellyfish populations is an interesting field of study on its own, another particularly intriguing question is whether jellies are increasing globally. Jellyfish bloom as a natural part of their life cycle, but they are also able to exploit disturbed ecosystems faster than most other species. The trick, of course, is being able to tell one pattern from the other.

Compelling evidence suggests that jellyfish blooms vacillate on a 20-year cycle. So on the one hand, this would seem to argue against an increase over time. On the other hand, numerous studies indicate that jellyfish populations are increasing in habitats impacted by overfishing, pollution, climate change, and other destabilizing effects, and that our growing human population is likely to progressively disturb natural habitats. Mounting evidence demonstrates that jellyfish can and do flourish where other species falter, and in some cases may be a visual indicator of an ocean out of balance. We delve more into this compelling subject in chapter five, where we look at the mechanisms and effects of these blooms.

**Far left** Under certain conditions a burst of nutrients may trigger a bloom of salps so thick that visibility may be reduced to almost nil. Salps don't sting, but can feel strangely disturbing when they hit swimmers' skin.

**Left** Some species, like the moon jellyfish (*Aurelia* spp.) have two occurrence patterns simultaneously. They bloom during the warmer months of the year, and each day they rise to the surface en masse in apparent spawning aggregations.

# THE PLEUSTONIC COMMUNITY

One group of marine organisms, known as the pleuston, thrives in the harshest of all oceanic environments. The pleuston comprises those organisms that live at the air-water interface. All sorts of creatures—jellyfish, crustaceans, mollusks, worms, and others—are adapted to live here, and they form their own community with surprising ecological interactions.

### Challenges of the Air-water Interface

Pleustonic species have to deal with both aquatic and aerial forces. Their delicate bodies are vulnerable to rapid drying when exposed to air and sunlight. Moreover, air temperature and water temperature fluctuate at different rates and to different extremes over the course of a 24-hour period; this means that different parts of an animal might be varying temperatures at the same time, creating potential problems for metabolic function or enzyme action. These creatures also must deal with predators from above and below, while being unable to fully retreat to either realm for safety or explore either zone to find food.

Most of the pleustonic organisms are an intensely blue color, a shade midway between teal and fluorescent blue. It is believed that this color helps protect these creatures from damaging ultraviolet light, and it may also help camouflage them against the ocean from aerial predators.

### The Sailing Jellies

Some of the most marvelous of all species occur in the pleustonic environment. The Portuguese Man-of-war and the Blue Bottle (*Physalia* species, page 34) reside here, with their bubble-like float sticking up in the air and their tentacles streaming in the water below. The Blue Button (*Porpita porpita*, page 78) lives here as well, with its disk of concentric circles sitting on the surface and its brilliant blue fringe of tentacles just underneath. The By-the-wind Sailor

(*Velella velella*, page 78), *Porpita*'s cousin, is found here too, with its oblong blue raft and silvery vertical sail. These creatures pass their lives traveling around the vast oceanic surface in large armadas.

In *Physalia* and *Velella* populations, different members ride different winds. Species of these two genera develop in "right-handed" and "left-handed" forms, with their sails oriented to one side of the body or the other. When the wind picks up and drives these creatures toward shore—and to their deaths—only those with one orientation will be blown, depending on the wind direction; the others remain behind to continue propagating the species.

## Other Pleustonic Creatures

Another peculiar member of the pleuston is not a jellyfish at all, but it preys on jellies—and stings like one. The Blue Dragon or Sea Lizard (*Glaucus atlanticus*, along with several kin), a stunningly strange, blue-, black-, and silver-striped creature, is a nudibranch, or sea slug. It has two pairs of limb-like projections, each bearing numerous fingerlike appendages. Each finger stores undischarged stinging cells obtained from its jellyfish prey, which may be deployed as needed for defense. The Blue Dragon keeps afloat by gulping air, which pins it to the surface. It moves by using its sluglike muscular foot to glide along the underside of the surface: remarkably, instead of crawling on the top of the water, it literally crawls on the bottom of the air.

Another common member of the pleuston is the Violet Bubble Snail (*Janthina janthina*). *Janthina* is a small hard-shelled mollusk that, like *Glaucus atlanticus*, eats pleustonic jellies. *Janthina* stays afloat by creating and clinging to a raft of air bubbles trapped in mucous. If it makes one wrong move, it falls away to meet a certain death, as it is unable to refloat itself. *Janthina*'s shell—with its beautiful shades of lilac and lavender—displays a common marine coloration pattern called countershading, which helps camouflage aquatic animals from visual predators (or prey) above and below. Usually animals with this pattern have a darker hue on top and a lighter hue on the bottom. Intriguingly, in this upside-down-hanging snail, the usual pattern is reversed: the shell is darker on the bottom, which faces up when the snail is hanging on to the underside of the bubble raft, and lighter on the top, which projects downward.

These pleustonic creatures, and others too, can be observed at any tropical or temperate shoreline of the world following an onshore breeze. Many people stay away from the beaches at these times for fear of being stung by jellyfish, but the keen observer can find the amazing creatures of the pleuston lurking then.

**Left** Pleustonic animals form a strange community, semi-aquatic and semi-aerial. The Sea Lizard (*Glaucus atlanticus*) eats jellyfish like *Porpita*, sequestering stinging cells into its "fingers" for its own use.

**Above** From time to time, millions of By-the-wind Sailors (*Velella velella*) are blown ashore, turning entire stretches of coastline purplish blue. Underneath the disk and sail lies a teeming polyp colony.

# PELAGIC ECOLOGY

Most of the open ocean is a dark and mysterious place. Sunlight penetrates only a fraction of the depth; this region is the epipelagic zone. The remainder is permanently aphotic, or lightless. Shrouded under layers of water, these realms—the mesopelagic, bathypelagic, and abyssopelagic zones—are all but impenetrable to humans, isolated by vast distances, both horizontal and vertical. The deep ocean is prohibitively expensive to access, and scientific expeditions rely on massive nets deployed from huge ships. Most marine biologists never get far below the surface; for the few that do, the journey is often a life-changing experience. The creatures that live in the open ocean, especially the depths, often defy imagination.

## The Deep Ocean

The deep pelagic zones are home to an incredible diversity of jellyfish. This is their domain. Darkness allows these tactile predators to have the advantage over visual predators. Their watery composition allows them to hang motionless, like clouds in the sky, without expending extra energy to keep their position. They simply sit and wait for unsuspecting prey to happen by. Their low energetic needs allow them to go a long time between meals or matings, a distinct advantage in an environment whose vastness keeps predator, prey, and partner apart.

This ecosystem is the most homogeneous on Earth. It is water, just water. The creatures that live here spend their whole lives drifting, never touching bottom. The temperature throughout most of the open ocean is a fairly constant 4°C (39°F), about the same as a home refrigerator. The pressure is enormous—our own lungs would implode within several hundred feet of the surface—but jellies' high water content makes their bodies uncollapsible.

## Finding Others in the Open Ocean

Jellyfish spend all their time seeking food, escaping predators, or finding a mate. These activities govern their whole existence and are more difficult in the open ocean than anywhere else on Earth because of several factors.

First, the biodiversity and biomass are lower, so the potential for finding food or fellowship is reduced. Second, the viscosity, or thickness, of the water and the lack of a substrate to push off from makes escape as challenging as it might be in outer space. Indeed, this is inner space, far removed from that thin margin in between where we live so comfortably.

Jellyfish interact with numerous other types of organisms. In shallower coastal waters, many of these interactions are mutually beneficial for both the jelly and the associate. One example is the symbiotic algae found in blubber jellyfish. However, in the open ocean the relationships between jellies and other species generally come down to two types: predator-prey and host-parasite. In these waters, jellies are frequently parasitized by small insect-like crustaceans or other jellies.

The food-finding strategies jellyfish have adapted in open water are fascinating. Some types of jellies will set their tentacles in a spiral pattern, creating an impenetrable curtain for their prey. Other species array their tentacles outward, creating a big field around their bodies. Still other species drift or swim slowly, trailing their tentacles behind, hoping to catch tiny plankton in their turbulence.

The Flying Saucer Jelly (*Atolla*, page 168) uses a specialized tentacle for luring its siphonophore prey. Others species, including some siphonophores, use their tentacles like jigs.

## Mimicry

Mimicry is common in the ocean, both in jellyfish that imitate other species and vice versa. One example is in the siphonophores, where some species possess batteries of stinging cells that resemble copepods or fish larvae, common prey items in the pelagic environment. When the siphonophore trolls for food, these structures are used as lures; they even contract intermittently, emulating the jerky movements of the prey species they mimic. Siphonophores use these lures to attract fish that prey on copepods and fish larvae.

Certain species of cephalopods will mimic jellies at times. The Mimic Octopus (*Thaumoctopus mimicus*) of Indonesia and the larvae of some squids react to disturbances or threats by spreading out their tentacles or curling the body up in a way that creates the impression of certain jellyfish species.

## Reproduction

The diversity of hunting and defensive strategies jellyfish employ in the open ocean is paralleled by the array of reproductive strategies, some of which require the services of other species. Although most open-ocean jellies do not have polyps, the polyps of those that do often grow on other species. Some hydroids, for example, grow on sea butterflies (swimming mollusks of the suborder Thecosomata) or fish. Others grow parasitically on other jellies. Some species of the hydrozoan narcomedusae, for example, have parasitic juvenile medusae that live inside the bodies of other narcomedusae or even other types of jellyfish.

The creatures that live in the vast expanses of the pelagic zone, where there is nowhere to hide, have adapted ways to stay competitive, often at the expense of others.

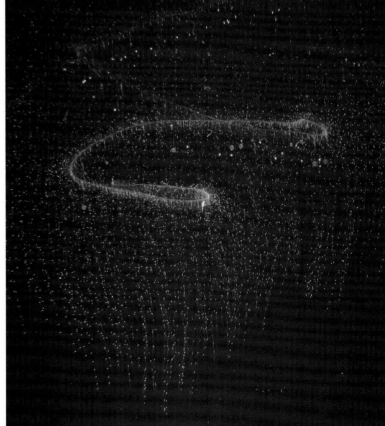

**Above right** The Mimic Octopus (*Thaumoctopus mimicus*) can fool predators by changing its body shape, color, and behavior to mimic a diverse range of species including crabs, lion fish, sea snakes, and jellyfish.

**Right** Siphonophores deploy thousands of tentacles for food capture. This stunning deep-sea species was observed using the CSIRO's Profiling Lagrangian Acoustic Optical System (PLAOS) during a voyage of exploration in the Great Australian Bight.

# MIGRATION

Jellyfish are drifters, able to orient up or down in the water column but, with few exceptions, unable to fight a current. Indeed, they are part of the plankton, a community that by definition drifts rather than swims. In spite of this passive lifestyle, many jellyfish are able to undergo complex horizontal and vertical migrations.

**Vertical Migration**

Many organisms of the deep sea, including jellyfish, migrate vertically up to a mile every day. They spend the daylight hours at depth and then begin ascending around dusk. Many reach the surface, or at least come close, before commencing their descent back to the depths before dawn. The reasons for this migration are varied. Some species ascend because their prey lives in the sunlit zone (the epipelagic), but being there during daytime would make the jellies too vulnerable to predation. Other species, however, probably simply follow their migrating prey, which may travel for a similar variety of reasons.

Migration may not be all about escaping predation or following prey. It may be energetically more advantageous to eat at the surface and digest at depth than to stay at just one location or the other. Prey are far more plentiful near the surface than in deeper water. However, the colder water at depth means that an organism's metabolism slows down

considerably, so it conserves energy by spending time there. The animal can therefore expend a higher percentage of its total energy on digestion and reproduction while in deeper water, whereas in shallower water it expends more energy on respiration, general life support, and predator evasion.

A lengthy migration would certainly require a huge expenditure of energy if an organism relied entirely on swimming. In at least some species, chemistry does the work; these jellyfish rely on the chemical properties of ions (atoms with positive or negative charges, some heavier than others) to adjust their buoyancy. Jellyfish may actively exclude heavy ions such as sulfate from their own cells, or accumulate lighter ions such as ammonium from the surrounding seawater. Even the slightest change is enough to accomplish an effortless migration. When the time comes to sink back down, they just reverse the process. Not all species, however, appear to exchange ions for migration; others simply do it the old-fashioned way—they swim.

**Left** Millions of Spotted Jellyfish (*Mastigias papua*) migrate each day around the lakes of Palau, following the sun to maximize exposure for their algal symbionts. Similar sun-compass migrations are known for other species as well, but for unknown reasons.

**Above right** In both the deep sea and shallow coastal regions alike, many types of jellyfish undergo daily vertical migrations. In the deep sea, species typically rise to the surface at night and sink back to deeper waters during the day. In contrast, shallow species often sink down at night and come to the surface once or twice during daylight hours.

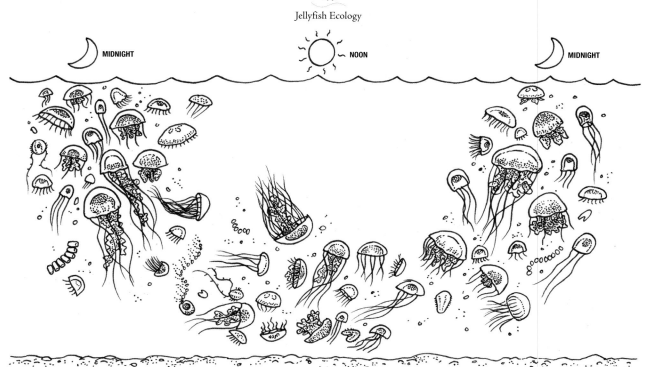

MIDNIGHT

NOON

MIDNIGHT

**Shallow-water Migration**

While vertical migration is common in deep-sea species, some coastal species also migrate vertically, albeit over a much shallower distance. Many populations of moon jellies (*Aurelia* spp., page 130) from around the world migrate to the surface once or twice each day in brief spawning aggregations, and then disperse again after about half an hour. Industries that rely on water flow or intake from the ocean are particularly vulnerable to invasion by these ephemeral aggregations.

At least one species of *Aurelia* in western Canada also performs daily horizontal migrations across the inlet in which it lives, using the sun as a compass. These medusae are distributed randomly when the sky is overcast and at night, but on sunny days the medusae migrate in a southeasterly direction and form enormous masses along the inlet's southeastern shore, the portion of the coastline that gets the longest afternoon sun exposure. Similarly, the Spotted Jelly (*Mastigias papua*, page 160) in Palau's inland salt lakes forms temporary aggregations each day; the different populations cluster in response to the specific conditions in their home lake.

**Migration Mysteries**

Two particularly intriguing migration mysteries await resolution. Australia's endemic Deadly Box Jellyfish (*Chironex fleckeri*, page 50) and Common Irukandji (*Carukia barnesi*, page 154)—two of the world's most venomous animals—appear to undergo seasonal migrations that often bring them into contact with swimmers. *Chironex* juveniles are observed coming out of rivers following the first heavy rain of the wet season each year; however, the species has never been observed migrating back into the rivers. In fact, its polyp stages have been found only once in a river, so its relationship to rivers remains a curiosity.

*Carukia barnesi*'s migration is even less clear. Typically *Carukia* infests certain beaches for one or two days. These infestation events have been strongly correlated with certain wind patterns, making them largely predictable. Historically it has been believed that the jellyfish migrate horizontally from offshore into shallow waters. However, medusae ranging from hours old to weeks old typically occur together, raising the prospect that they are migrating up from the bottom rather than in from deeper water.

# IRUKANDJI JELLYFISH

BALMY TROPICAL AND SUBTROPICAL beaches are a preferred destination for millions of people every year—the Great Barrier Reef, Hawaii, Florida, the Caribbean, Thailand, and countless Indo-Pacific islands. But there is an invisible danger that lurks in the crystal-clear waters: Irukandji. The name is of Australian Aboriginal origin but it now applies to more than a dozen species, all of the order Carybdeida. These species cause excruciating, life-threatening stings.

## Irukandji Syndrome

The initial sting of an Irukandji jellyfish is often so mild it escapes notice. But after a period of five to 40 minutes (depending on the species), the constellation of symptoms known as Irukandji syndrome begins: severe lower back pain, difficulty breathing, profuse sweating, cramps and spasms, coughing, a creepy skin feeling, and a sense of impending doom. In some cases, severe hypertension (high blood pressure) may cause pulmonary edema (fluid on the lungs), heart failure, or stroke.

## *Carukia barnesi* and Other Species

The first-named Irukandji species—and the best known—is *Carukia barnesi*, the Common Irukandji. Its four tentacles are 100 times as long as its centimeter-long (half-inch) body and as fine as cobwebs. Long after scientists learned that the mysterious Irukandji illness came from *Carukia*'s stings, the predictability of its swarms along Australian beaches remained elusive. However, the forecasting mystery was solved in 2012, when researchers pinpointed the subsidence of predominant trade winds as the primary factor in Irukandji infestations.

Numerous other species cause Irukandji syndrome. Species of *Malo* (page 200), *Morbakka*, *Gerongia*, and *Keesingia* are in the same family, the Carukiidae. *Alatina* species (family Alatinidae) are periodically common along Waikiki Beach in Hawaii, the Florida Keys, Grand Cayman in the Caribbean, and the Great Barrier Reef. Some hydrozoans and scyphozoans may also cause Irukandji syndrome, spanning from Wales and Boston to Melbourne and Cape Town.

**Scientific name** *Carukia barnesi* (and other spp.)

**Phylogeny** PHYLUM Cnidaria / CLASS Cubozoa / ORDER Carybdeida

**Notable anatomy** thimble-shaped to tube-shaped body, with four tentacles, each bearing numerous bands of stinging cells

**Position in water column** shallow coastal waters, tropical reefs and islands

**Size** bell 1 to 50 cm (0.4–20 in.) in height; tentacles to about 1 m (3 ft. 3 in.)

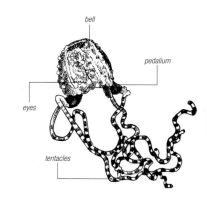

bell

pedalium

eyes

tentacles

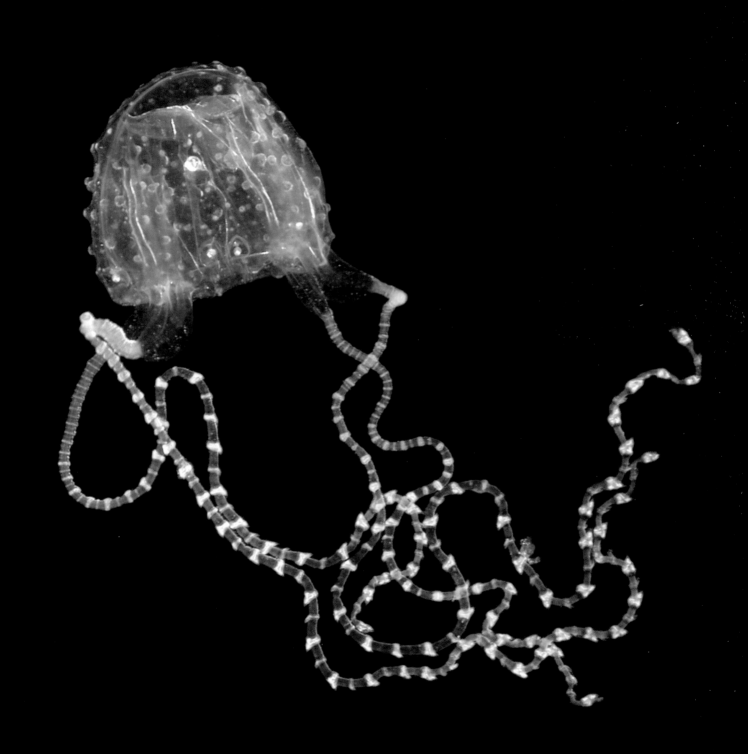

# HULA SKIRT SIPHONOPHORE

ONE OF THE MOST STUNNINGLY beautiful siphonophores is a creature with the name *Physophora hydrostatica*, which can be loosely translated as "bladder-bearing water skeleton." *Physophora* typically lives in the midwater zones of the open ocean but, much to the delight of divers and snorkelers, is occasionally brought to the surface by upwelling.

### Very Alluring
*Physophora* is hard to mistake for any other species. Its most noticeable feature is a skirt of sparkly pink-orange palpons, or stinging batons. Above that is a zone of bubble-like nectophores, or swimming bells, and above these lies a small, silvery gas-filled float, used for buoyancy. Several long, fine tentacles stream out from the bottom between the palpons. The whole colony may reach several inches in height, excluding the tentacles.

Some types of siphonophores have hundreds of tentacles that they deploy into long curtains for capturing unsuspecting prey, while others trawl their tentacles through the water to catch food items. *Physophora*'s tentacles are knobbed with structures that resemble tiny planktonic creatures, offering a clue as to how this siphonophore might find its prey. Some researchers have proposed that these knobs may serve as lures for visual predators.

### Skirt Flaring
*Physophora* usually drifts passively or coordinates the pulsations of its nectophores to swim slowly. At times, however, it suddenly flares its palpons up in a way that resembles a woman's skirt being blown upward by the wind. The reason for this behavior is unknown, but it may serve as a defensive action that makes the colony suddenly appear larger.

**Scientific name** *Physophora hydrostatica*

**Phylogeny** PHYLUM Cnidaria / CLASS Hydrozoa / ORDER Siphonophora / SUBORDER Physonectae

**Notable anatomy** long, narrow float atop zone of nectophores; skirt of pink-orange palpons; several long, fine tentacles streaming down

**Position in water column** epipelagic, mesopelagic, bathypelagic

**Size** colony to about 8–12 cm (3–5 in.) long

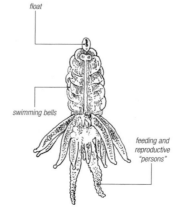

float

swimming bells

feeding and reproductive "persons"

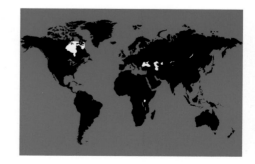

# AUSTRALIAN SPOTTED JELLYFISH

In 2000, fishermen and others along the coastlines of the Gulf of Mexico began noticing unusual numbers of unfamiliar jellyfish. These creatures were hard to miss: they were big, bulky animals, to about 50 centimeters (a foot and a half) across, and their conspicuous coloration pattern—whitish with bright white spots—stood out against the blue-green of the ocean.

Scientists quickly realized that these jellies belonged to *Phyllorhiza punctata*, a species native to southwestern Australia. It is surmised that they invaded the Gulf through the Caribbean, where they reportedly have been living for more than 50 years.

In its native habitat, *Phyllorhiza* is dark olive brown with white spots. The brown color comes from symbiotic algae living in the jelly's tissues; carbohydrates made by these algae help the jellyfish meet a substantial portion of its nutritional requirements. Curiously, however, the population in the Gulf of Mexico has lost its algal symbionts, making it appear whitish or bluish.

Because the invasive *Phyllorhiza* lacks symbiotic algae, it must eat plankton from the surrounding water. The large number of jellies means a lot of mouths eating great quantities of plankton and putting pressure on the local ecosystem.

## Unloved Pest

There is a sad irony to the ecology of these two populations. In Australia, *Phyllorhiza* is a troublesome pest, partly because its symbionts allow it to grow quickly and reproduce unchecked without having to find food. However, the American population is also a pest, in this case because it has to find food, so it eats other species out of house and home. Either way, the jellies win.

---

**Scientific name** *Phyllorhiza punctata*

**Phylogeny** PHYLUM Cnidaria / CLASS Scyphozoa / ORDER Rhizostomeae

**Notable anatomy** large, bulky body with eight frilly oral arms, each bearing a club; Australian population dark olive brown with white spots, American population translucent whitish with bright white spots

**Position in water column** epipelagic, in neritic zone

**Size** bell to 50 cm (20 in.) in diameter

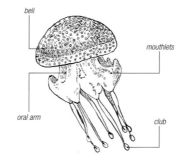

bell

mouthlets

oral arm

club

# SPOTTED JELLY

IMAGINE DRIFTING WEIGHTLESSLY, silently, mesmerizingly through a blizzard of harmless polka-dotted jellyfish, as if you were inside a living lava lamp. This is not a dream scenario; in a couple of salt lakes, far in the western Pacific in the tiny island nation of Palau, this experience has become a major tourism draw over the last few decades.

The Spotted Jelly (*Mastigias papua*), also known as the Golden or Lagoon Jelly, is yellowish to pinkish, scattered with conspicuous white spots. Its body is globular, and it has eight short, downward-hanging oral arms. At maturity, the diameter of the bell reaches only about eight centimeters (three inches). But what it lacks in size, it makes up for in numbers: some estimates put the number of jellyfish in the lakes at more than 1.5 million. The species is also found in coastal waters throughout the Indo-Pacific.

## Blubber Jellies

*Mastigias* is a rhizostome, or blubber jelly. It gets its nutrition from symbiotic photosynthetic algae living in its tissues. Those confined to inland lakes have almost no predators and no need to defend themselves—unless they get too close to the bottom, where they may be eaten by a hungry sea anemone. Nonetheless, they have no ability to sting.

Rhizostomes, like most cnidarian jellyfish, have a polyp stage. They generate medusae by strobilation (described in "Scyphozoan Life History," pages 62–63), but often produce only one ephyra, or baby jellyfish, at a time. The trade-off to producing only a single medusa is that the polyp can recycle faster between strobilation events, keeping the process going more or less continuously, rather than waiting for one big burst when the conditions are right.

**Scientific name** *Mastigias papua*

**Phylogeny** PHYLUM Cnidaria / CLASS Scyphozoa / ORDER Rhizostomeae

**Notable anatomy** globular body with eight short, thick oral arms, each bearing a thick club; golden to pinkish with white spots

**Position in water column** surface waters in inland saline lakes or coastal waters

**Size** bell to 8 cm (3 in.) in diameter

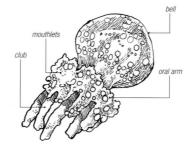

# FRESHWATER JELLYFISH

MOST PEOPLE HAVE NEVER HEARD of freshwater jellyfish, and those fortunate enough to encounter them are typically surprised and delighted. One species, *Craspedacusta sowerbii*, is about two centimeters (less than an inch) across, and its body is delicate and lacy. It is quite a sight to behold in bloom conditions; snorkelers, lying motionless in the water and watching through a mask, get the impression of a blizzard of snowflakes drifting by.

## Hitchhikers

Native to China, *Craspedacusta* is now commonly found in lakes, streams, man-made reservoirs, and backyard lily ponds around the world. It is thought that the species spreads when its tiny polyps, clinging to bits of mud on birds' feet, hitch rides between water bodies. The species may also, in some cases, be transported via the ornamental aquatic plant trade.

Most infestations of *Craspedacusta* are either all male or all female but not both. This is likely the result when only a single polyp is transferred to a body of water. When that polyp clones, making many replicates of itself, its offspring can release enough medusae to create a swarm. But without both sexes, *Craspedacusta* is unable to maintain a population over time, and eventually dies out.

## Mosquito Eater

The world-traveling *Craspedacusta* may prove useful to humanity. It eats mosquito larvae. Whether this aspect of its ecology may eventually lead to a natural form of mosquito control is not yet clear. Introducing alien species into virgin environments can be problematic, and time and again has proven disastrous the world over. In Africa, especially, other native species of freshwater jellyfish may prove to be a more appealing alternative.

**Scientific name** *Craspedacusta sowerbii*

**Phylogeny** PHYLUM Cnidaria / CLASS Hydrozoa / ORDER Limnomedusae

**Notable anatomy** small lacy body with many tentacles, a long manubrium, and four pocket-shaped reproductive organs

**Position in water column** all depths in shallow lakes, streams, reservoirs, and lily ponds

**Size** bell to about 2 cm (0.8 in.) in diameter

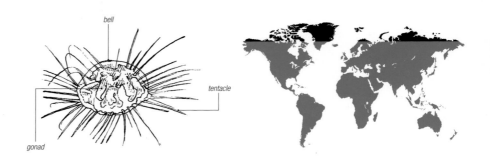

bell

tentacle

gonad

# JELLIES WITH BITE

People often claim to have been "bitten" by a jellyfish, but their terminology is erroneous. Jellyfish sting; they do not bite—at least most of them. As strange as it may seem, some jellyfish actually do have teeth. Comb jellyfish in the genus *Beroe* are simple, pocket-shaped creatures lacking tentacles, lobes, or other appendages. However, they have special highly modified, enlarged cilia lining the mouth that function as teeth.

## Feeding Habits

When *Beroe* finds a suitable creature to devour, it swims toward the prey, passes it, and then turns back for the kill. If the prey is small enough, *Beroe* wraps its lips around the victim and engulfs it, the way a snake pulls in a mouse.

If the prey is too big to fit in the mouth, *Beroe* bites off chunks until its belly is full. *Beroe* is not dangerous to people, however; it eats only gelatinous prey, mostly other ctenophores.

Being pocket-shaped presents all sorts of swimming challenges. *Beroe* swims with the mouth forward, running the risk of filling the body with water. But along with its teeth, *Beroe* has adhesive cells lining the mouth; these stick the lips together so the animal moves more efficiently as it swims.

Different species of *Beroe* are found in every ocean and at all depths. They are bioluminescent, defending themselves with flashes of brilliant blue along their comb rows. *Beroe* jellyfish are hermaphroditic—each individual is both male and female at the same time.

**Scientific name** *Beroe* spp.

**Phylogeny** PHYLUM Ctenophora / CLASS Nuda / ORDER Beroida

**Notable anatomy** pocket-shaped body, with eight longitudinal rows of cilia

**Position in water column** epipelagic, mesopelagic, and bathypelagic

**Size** body typically 6–10 cm (2.35–4 in.) long at maturity, maximum to 30 cm (12 in.) long; varies by species

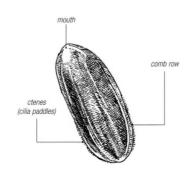

mouth

comb row

ctenes
(cilia paddles)

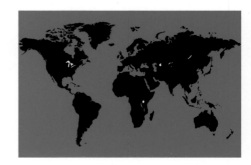

# SANTA'S HAT JELLY

Many creatures are bioluminescent, or able to make their own light, such as fireflies, glowworms, and several types of fish, crustaceans, squids, and, of course, jellyfish. Usually jellyfish employ bioluminescence as a flash to create a startle response. In some species it occurs as a glow. One type of jellyfish, however, puts its bioluminescence to different use.

## Twinkling and Camouflage

The deep-sea *Periphylla periphylla* twinkles. Tiny pinpricks of light flash on and off, creating a sparking appearance. It is possible that *Periphylla* uses this twinkling as camouflage, letting its scattered luminescence belie its true size or shape.

*Periphylla*, nicknamed the Santa's Hat Jelly, makes use of another type of camouflage as well. While a flash of light startles some creatures, others take the flash as a sign of a potential meal. Thus, any unshielded light from swallowed plankton prey, which the jellyfish often does not kill

immediately, could mean trouble for the jellyfish. *Periphylla* has an opaque red stomach and this effectively shields from view any bioluminescent responses that struggling prey in its stomach may give off. At depth, red looks black, making the stomach essentially invisible.

## Fjord Invasion

As wondrous as *Periphylla* is in its natural habitat—the depths of every ocean—it has become an unforeseen pest in Norwegian fjords, at all depths including surface waters. When fish populations began to dwindle in the fjords due to overfishing, *Periphylla* took up residence, a habitat surprisingly similar to its natural home—cold, dark, deep, and still. Tactile predators such as *Periphylla* do not need to see to eat, giving them the advantage over visual predators like fish. At least two fjords appear to have reached a new species balance, with *Periphylla* as the top predator.

**Scientific name** *Periphylla periphylla*

**Phylogeny** PHYLUM Cnidaria / CLASS Scyphozoa / ORDER Coronatae

**Notable anatomy** large conical bell with a dark red stomach inside

**Position in water column** epipelagic to abyssopelagic; also found in all depths in fjords of Norway

**Size** body to 30 cm (12 in.) in height

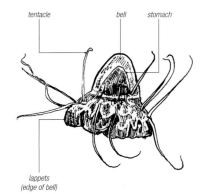

tentacle    bell    stomach

lappets
(edge of bell)

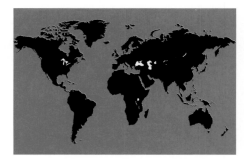

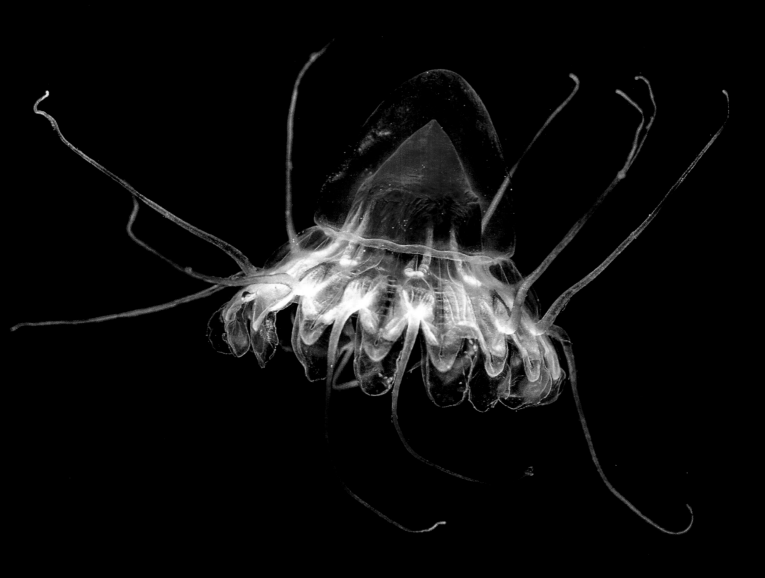

# FLYING SAUCER JELLYFISH

IMAGINE A JELLYFISH FLATTENED like a Frisbee, with a dome rising in the middle and radiating ripples along the outer margin. Give it about 20 tentacles around the edge, each alternating with a rhopalium, or sensory knob. Color the stomach deep red to mask any bioluminescent light emitted by struggling prey within. This describes the so-called flying saucer jellies of the genus *Atolla*.

Several species are known, all from the deep sea. The most common, *Atolla wyvillei*, grows quite large, up to about 15 centimeters (six inches) across, and is deep red throughout. Another, *Atolla clara*, is also quite large but lacks the full-body red pigment that makes *A. wyvillei* so striking. Smaller, more delicate varieties are also common.

Recently, thanks to submersibles that can travel to the depths of the ocean, scientists made an exciting discovery: *Atolla* uses its tentacles to partition its predation efforts.

When *Atolla* is observed in its natural habitat, one of its trailing tentacles is always longer than the rest. It uses this enlarged tentacle to capture *Nanomia*, a common genus of siphonophore; it potentially captures other gelatinous prey items as well. The other tentacles, it seems, are used to capture zooplankton prey, such as copepods. *Atolla* is prey in turn: at least one species of deep-sea shrimp—the big, red, spiky, tough-looking *Notostomus robustus*—has been observed feeding on it.

### Distinctive Bioluminescence

Like many deep-sea organisms, *Atolla* is capable of emitting bioluminescence. But this is no ordinary flash or twinkle; *Atolla*'s light is a racing wave of fluorescent blue traveling rapidly around the bell in a circular pattern.

**Scientific name** *Atolla* spp.

**Phylogeny** PHYLUM Cnidaria / CLASS Scyphozoa / ORDER Coronatae

**Notable anatomy** flattish, saucer-shaped body with a central dome

**Position in water column** mesopelagic, bathypelagic, in open ocean

**Size** body to about 15 cm (6 in.) in diameter; varies by species

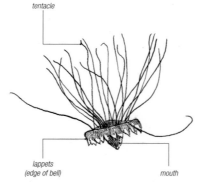

tentacle

lappets
(edge of bell)

mouth

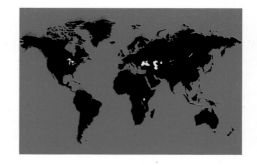

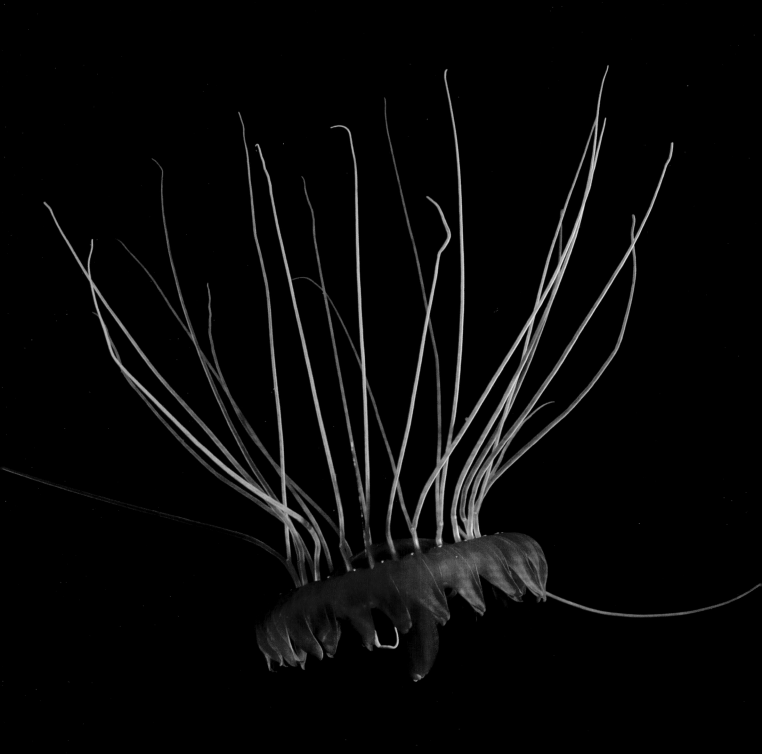

# LEAF-CRAWLER JELLIES

WHILE WE GENERALLY THINK of jellyfish as drifters swimming in the open ocean, there are a few types that are unable to swim. Among these are the stauromedusae (page 18), in which the medusa is flipped upside down, so it resembles a sea anemone, with the mouth and tentacles facing up. The leaf-crawler jellies of the genera *Staurocladia* and *Eleutheria*, on the other hand, live with the normal side up for a jellyfish—that is, mouth down—but lack the ability to swim. They spend their whole life crawling among algae and sea grasses on their tentacles.

The tentacles of these jellyfish are highly modified for an entirely benthic, bottom-dwelling existence. Each tentacle is bifurcated, with one of the two branches facing downward and ending with an adhesive sucker for crawling and sticking to leaves. The other branch faces upward into the water, where rings of stinging cells and a knob are used for prey capture. Prey consists of copepods and other small planktonic and crawling species. The stinging branch of each tentacle is waved in the direction of approaching prey, and then the knob is thrust toward the prey to make contact. Prey animals are often caught by their antenna and become paralyzed by the venom from the stinging cells within about 10 seconds.

## Reproduction

Species in the genera *Staurocladia* and *Eleutheria* are simultaneous hermaphrodites—both male and female at the same time—and they can self-fertilize. Larvae are brooded internally. They can also bud new medusa clones between their tentacles.

Mature medusae of these two genera reduce their feeding rate while they are brooding young, presumably as a way of reducing their overall size. A larger size would make them more vulnerable to becoming dislodged from their leaf or damaged otherwise.

**Scientific name** *Staurocladia* and *Eleutheria* spp.

**Phylogeny** PHYLUM Cnidaria / CLASS Hydrozoa / ORDER Anthoathecata

**Notable anatomy** minute flattened body with up to 60 fine, bifurcated tentacles, one branch with nematocyst clusters pointing upward, and the other with a downward-facing sucker

**Position in water column** benthic, on algae and sea grasses

**Size** body to about 2.5 mm (0.1 in.) in diameter at maturity; varies with species

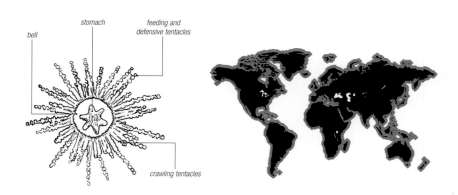

bell

stomach

feeding and defensive tentacles

crawling tentacles

# SPLENDID BELL JELLY

ONE OF THE MOST BEAUTIFUL of all jellyfish is the colorful, wild-looking *Polyorchis penicillatus*. Its deep bell-shaped body, many dozens of thin, graceful tentacles, and marginal row of red eyespots make it unmistakable. Decades ago, it was common in bays and harbors along the Pacific coast of North America; today, however, it seems to be vanishing. Whereas many jellyfish respond positively to degrading water quality, it appears that *Polyorchis* is falling victim to changes in the oceans.

### Death by Sea Slug
It has been known for many years that nudibranchs (sea slugs) eat hydroids and sequester undischarged stinging cells for use in their own defense. Medusae are generally thought to be out of range of these mollusks, because they can just swim away. However, at least one species of nudibranch—the Opalescent Sea Slug (*Hermissenda crassicornis*)—has worked out how to eat *Polyorchis penicillatus*. The slugs glide onto rafts of sea grass from rocks, wharf pilings, or other access points. When *Polyorchis* swims up toward the water's surface, which it does naturally, it gets stuck among the sea grass blades. *Hermissenda* responds to the vibrations given off by the struggling medusa, and comes racing over. Reaching the jellyfish, the slug rears up and pounces on a tentacle, severing and ingesting it. Then it rears up and pounces on another, and so on, until it has removed all the tentacles. This type of attack is lethal to *Polyorchis*, though it remains unclear whether the inability to get out of the grasses or the inability to catch prey is the main factor leading to death.

**Scientific name** *Polyorchis penicillatus*

**Phylogeny** PHYLUM Cnidaria / CLASS Hydrozoa / ORDER Anthoathecata

**Notable anatomy** deep bell-shaped body with a row of red eyespots around the margin and 100-plus long, fine tentacles

**Position in water column** shallow coastal waters, mainly in bays and harbors

**Size** body to about 5 cm (2 in.) in height

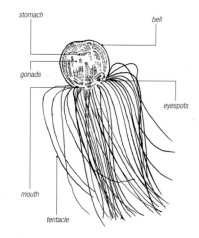

stomach · bell · gonads · eyespots · mouth · tentacle

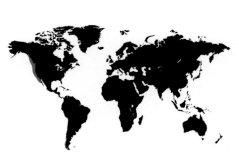

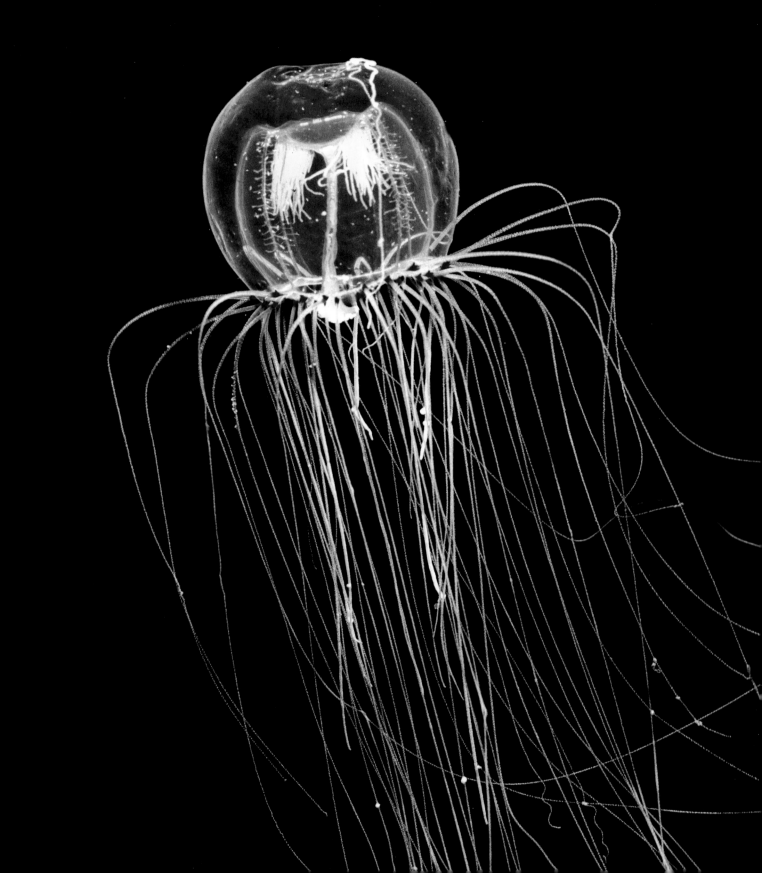

## CHAPTER FIVE

# OUR RELATIONSHIP WITH JELLYFISH

# HUMANS AND JELLYFISH

Most people do not give much thought to jellyfish. To many, they are just slimy creatures that sting. To others, they are a crunchy-chewy delicacy served in a savory sauce. To a few of us, they are beautiful, as mesmerizing as a lava lamp. Jellyfish are all these things, and they are much more. They date back to the dawn of animal life and they may hold the key to immortality.

THEY CAN KILL A HEALTHY ADULT in two minutes flat, or cause pain so bad you only wish you were dead. But far and away, the jellyfish topic dominating the media these days is their blooms. Whether they are invading beaches and stinging beachgoers, wiping out salmon farms, frightening off whale sharks, clogging power plants, forcing emergency nuclear-plant shutdowns, capsizing fishing boats, or taking over fjords, jellyfish are getting our attention.

## Scientific Research

As a scientific discipline, jellyfish blooms research has come about only within the last 20 years or so. The first publication on the subject, which appeared in 1995, essentially strung together a series of vignettes about shocking blooms around the world and hinted that there might be a problem.

Now, two decades later, many additional bloom incidents have been reported, and we have far more data than we had back then. How we interpret jellyfish blooms, however, is an issue of heated debate. One study concluded that there is insufficient evidence to support the contention, made by some scientists, that blooms are on the rise globally. After a follow-up study, the same researchers concluded that jellyfish operate on a 20-year bloom cycle, and claimed this explained the apparent increase. Yet a further study by the same research group found that jellyfish are on the rise in disturbed areas. Meanwhile, other studies have demonstrated, with compelling evidence, that jellies are indeed undergoing dramatic increases in disturbed areas, and that disturbed areas are increasing globally.

Jellyfish bloom science faces two quite different questions, one theoretical and the other practical, and both need research attention. The question of whether there has been a global increase in jellyfish has been thwarted by lack of robust data in most localities. However, studies on certain species in disturbed areas have shown a significant increasing trend in jellyfish biomass. It may be many decades before we have the long-term data that most scientists consider necessary for reliably testing global theoretical questions.

The second problem we face with jellyfish is our need for management solutions for those industries that are being impacted by blooms now. This problem is urgent and exists separately from the philosophical issue and regardless of any future trend. Almost every industry involved in marine activities is reporting increasing problems with jellyfish and huge labor and financial costs in managing them.

## A Perfect World for Jellyfish

Jellyfish may be thought of as either the perfect "fish" or the most tenacious of pests. Most fisheries are limited by brood stock and growth rate, but jellyfish offer a fast-growing renewable resource unmatched by conventional fish. The flip side is that these very same qualities make them superior pests unlikely to go away any time soon.

In previous chapters, we have looked at some of the particulars of the underlying biology and ecology that enable jellyfish to be such successful pests. In this chapter we turn our attention to the human, or anthropogenic, reasons for these problems—in other words, the things we humans are doing to make the world a bit more perfect for jellyfish.

WHAT CAUSES A BLOOM?

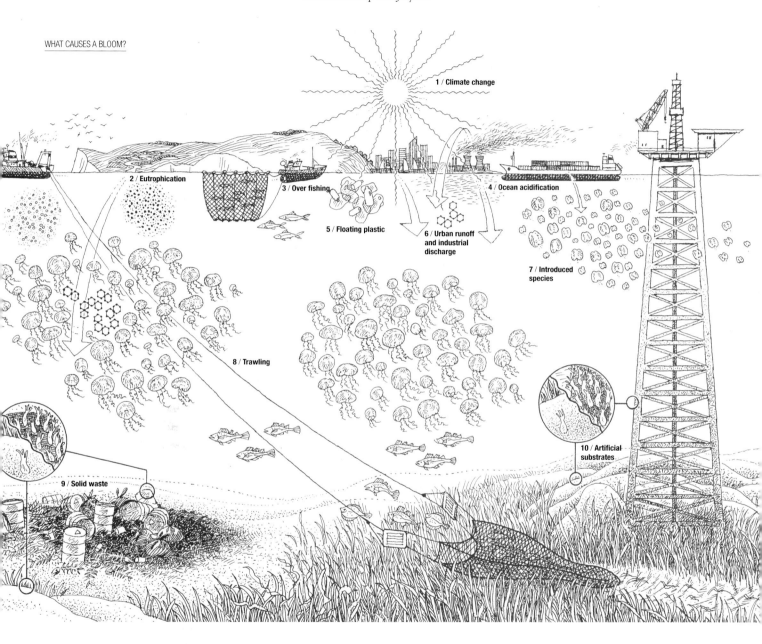

1 / Climate change

2 / Eutrophication

3 / Over fishing

5 / Floating plastic

6 / Urban runoff and industrial discharge

4 / Ocean acidification

7 / Introduced species

8 / Trawling

10 / Artificial substrates

9 / Solid waste

**Above  1.** Warming water stimulates reproduction, lengthens the growing season, and holds less dissolved oxygen than cooler water, making it harder for fish to survive. **2.** Nutrients from farm fertilizers and urban gardens stimulate phytoplankton and zooplankton, which jellyfish feed on. **3.** Fewer fish means more plankton for jellyfish. **4.** Carbon dioxide is absorbed by the ocean, lowering its pH and causing shells or skeletons of some species to dissolve away like osteoporosis, while the timing of the biological processes of others may be skewed. **5.** Jellyfish polyps get transported long distances on plastics, which disrupt the ecosystem when consumed by seabirds, fish, turtles, and marine mammals, benefiting jellyfish.

**6.** Toxic chemicals make their way into the food chain with catastrophic effects; jellyfish are often the last to be affected. **7.** Introduced species often have a deleterious effect on local ecosystems as they typically have no natural predators. **8.** Trawling flattens habitats that support healthy ecosystems and bycatch kills vast amounts of living things compared to the target species; over-turned rocks and even dead animals and plants provide new habitats for polyps. **9.** Pollution disrupts ecosystem function, which may leave jellyfish as the "last man standing." **10.** Some jellyfish polyps prefer artificial surfaces such as ports and marinas, oil and gas platforms, and aquaculture facilities.

# OVERFISHING

Overfishing means more than the obvious (taking too many fish); it encompasses any harvesting practice that limits the ability of fish to replenish their populations year after year. This is why, with many types of fishing, the catch may stay within the legal limits but nonetheless contribute to the progressive decline of a population.

Trawling, or fishing by dragging a large open net that gathers everything in its path, is one of the most destructive means of fishing. Up to 90 percent of each haul may be made up of nontarget species, which are often crushed, impaled, or decapitated in the process. Some people believe that these dead or dying discards are not a problem because the bycatch becomes food for other species. However, trawling rips up the sponges, corals, and other reef-builders that form the three-dimensional habitats that act as nurseries for vast numbers of fish, crustaceans, and mollusks. The practice can tip the scales in favor of weedy species by repeatedly disturbing the seabed and keeping it from regenerating. Moreover, killing these organisms limits the production of larvae that would have fed many of our target fish species in the future.

Another, often-ignored aspect of overfishing is a result of our human desire for the biggest and most ideal-looking fish. When we selectively harvest the fish we prefer, we leave the ones we have rejected as the breeders for future generations. Moreover, for many types of fish, the big, older, fatter females are the better breeders, generating larger numbers of bigger, healthier eggs. Over time, selectively taking out the larger fish results in diminishing populations of smaller-growing fish.

### Why Should We Care about Overfishing?
We hear the term "sustainable" all the time these days, but its meaning is not always clear. The phrase "sustainably caught fish" should indicate that the species is not in peril and that the fishing practices are not destructive.

Unfortunately, this is often not the case. A 2002 study by the United Nations found that 72 percent of the world's marine fish stocks were declining. A 2006 modeling study projected the collapse of fisheries by 2048. And a 2016 study found that global fish catches have been under-reported by a third: we are even worse than originally thought. Our growing population, it seems, is racing toward a crisis.

Fish farming is no more sustainable for the environment than harvesting wild fish (in fact, often less so). Shrimp farms are among the most polluting of all aquacultural enterprises, using unimaginable quantities of unthinkable chemicals to keep the stock alive and then flushing these toxic effluents into coastal waters. Salmon farms consume and waste vast

quantities of other fish, and only a tiny fraction of that food energy is converted to salmon flesh; these farms also often promote algal blooms and jellyfish blooms, and create anoxic (oxygen-free) dead zones under the cages.

Some people may find life without fish perfectly acceptable when the catch runs out, but more than 100 million people globally rely on fish as their primary source of protein. Furthermore, fish are also primary components of aquacultural feed, agricultural fertilizers, and health supplements for joints and brain function. Moreover, many other species that we care about also rely on fish, such as whales, dolphins, seabirds, bears, and most large fish, including sharks.

## What Does This Have to Do with Jellyfish?

Only recently have scientists come to understand the delicate ecological balance between fish and jellyfish. In a normal, healthy ecosystem, fish and jellyfish are each other's predators and competitors. All things being equal, fish are superior competitors because they are smarter and faster. But as fish decrease, more food is available for jellyfish, and as jellyfish increase, they eat more food, and it becomes harder for fish to find a meal. This creates a positive feedback loop that favors the jellyfish.

The mechanism by which jellyfish can drive down a fish population is devastatingly simple and effective. Jellyfish eat the eggs and larvae of fish, and they eat the plankton that the larvae rely on. The double impact of predation and competition can quickly shift the balance in an ecosystem from fish to jellyfish. And once jellyfish have assumed the role of top predator, as we discussed in "Predation and Competition" (pages 138–39), this ordering of the ecosystem appears to be incredibly resilient to change.

**Below left** Enormous aquaculture facilities in Xiapu County, Fujian Province, China, significantly alter coastal habitats, contribute excess nutrients leading to eutrophication, and offer millions of settling places for jellyfish polyps. Pest species find this type of environment particularly nurturing.

**Below right** Nomura's Giant Jellyfish (*Nemopilema nomurai*, page 194) flow by their billions into the Sea of Japan from China, crippling the Japanese fishing industry. Overfishing, trawling, coastal construction, and nutrient overload in China have contributed to these massive blooms.

# CLIMATE CHANGE

Despite all the media attention to climate change, many people are still unsure about what it actually is and is not. In its simplest sense, weather is what happens, whereas climate is what we expect to happen. Climate change, therefore, is a shift in the weather that we expect. More to the point, it is a change in the Earth's climate that includes both progressive warming and an increase in the frequency and amplitude of erratic weather. This means stronger storms more often, longer droughts, higher floods, earlier onset of spring, and so forth.

SUBTLE SHIFTS in the planet's mean temperature (too subtle for us to feel) create a cascade of atmospheric and ecological effects. Plants and animals perceive spring sooner and summer for longer. The lengthening of the summer season means more drying and less rainfall in temperate areas, which may result in more frequent or hotter forest fires. In the tropics, where cyclones and hurricanes gain their strength from warm surface water, the frequency of more severe events is on the rise. In the ocean, because warmer water holds less oxygen than cooler water, species with heavier respiration (such as fish and crustaceans) have to expend more energy to breathe than species with lower respiration (jellyfish and worms), shifting the balance over time from complex species toward those with simpler needs.

## Why Should We Care about Climate Change?

The popular media often portrays climate change as an issue of rising sea levels. However, although a rising sea threatens the continued existence of low-lying island nations such as Kiribati and the Maldives, it is not the biggest, most pressing threat for most of the people on Earth.

For most of us, food supply is a much bigger concern. While the threats to food security are myriad, three in particular are conspicuous. First, groundwater supplies (aquifers) are already in peril the world over, which makes us increasingly reliant on rainfall. As farmlands dry with diminishing water supply, we see an impact in the choices of food items that can be grown as well as the quantity and quality of the harvest.

Second, as coastal waters warm and hold less oxygen, they will support fewer fish. Fewer fish means less choice, higher prices, and diminishing resources for aquacultural feed and agricultural fertilizer.

Third, shifts in phenology, or the timing of biological events, may prove to be the most damaging outcome of all. When males and females of a species spawn at different times, reproduction fails. When predator and prey mature at different times, the predator starves, which in turn jeopardizes the food supply of its own predators. When plants and pollinators mature on different schedules, not only flowers are in peril but also most of our food crops. We will not notice these changes in most species until their populations are in free fall, and by then it may well be too late.

## What Does This Have to Do with Jellyfish?

Jellyfish contribute to two feedback loops with regard to climate change. First, warmer water revs up their metabolism and makes them grow faster, eat more, breed more, and live longer—all the things we do not want a pest to do. Jellyfish of larger populations, in turn, have the potential to contribute further to climate change. Their mucus and dissolved organic matter (what jellyfish researchers call

*EFFECT OF TEMPERATURE ON FISH AND JELLYFISH*

Cool water ◄┄┄┄┄┄┄┄┄┄┄┄┄┄┄┄┄┄┄┄┄┄┄┄┄┄┄┄┄┄┄► Warm water

"goo and poo") both attract a type of bacteria that shunts its energy away from the food chain and toward carbon dioxide production instead. Therefore, throughout their lives, and particularly when they die, jellyfish act as carbon dioxide factories.

The other feedback loop is based on predation and competition, as we see with other environmental stressors, such as overfishing and pollution. As jellyfish flourish they can outcompete fish and other species simply with their sheer numbers, emerging as the top predator.

**Above** Subtle temperature changes may help shift the balance from fish to jellies by having quite opposite effects on each. Even slight warming revs up jellies' metabolism, making them grow faster, eat more, reproduce more, and live longer.

In contrast, warmer water holds less dissolved oxygen, making it more difficult for organisms like fish to extract oxygen from the water. As fish struggle and jellies bloom, this creates a positive feedback loop, particularly in combination with other disturbances.

# OCEAN ACIDIFICATION

Ocean acidification has been called the "evil twin" of climate change. The ocean acts like a big sponge, absorbing carbon dioxide from the atmosphere. Once absorbed, carbon dioxide causes changes in the chemistry of seawater, making it become more acidic. The ocean is not turning to acid, but its pH—a number on a scale that measures how acidic or alkaline (or basic) a substance is—has been getting lower and shifting toward a less alkaline, more acidic state.

MARINE ANIMALS AND PLANTS extract the chemical calcium carbonate from seawater to make their calcified hard parts. But as the ocean absorbs more carbon dioxide, and the pH lowers toward the acidic side of the scale, the equilibrium of calcium carbonate is shifting, which is making it dissolve more readily in seawater. While boaters and beachgoers may not perceive any difference in the seawater, to organisms living in the water, the change is significant.

As oceans acidify, three things are happening to marine organisms, affecting them simultaneously. First, the shells and skeletons of some creatures are simply dissolving away, in a process somewhat similar to that of osteoporosis. The hard parts of corals and sea butterflies (mollusks of the division Pteropoda) are being disintegrated by the more acidic seawater, one molecule at a time. The shells and skeletons of these organisms are made of aragonite, a less stable form of calcium carbonate than the calcite that crabs, lobsters, fish, snails, clams, and sea urchins are made of. Without the hard shells and skeletons to protect and support their delicate bodies, these corals and pteropods will die.

Second, while some organisms are dissolving away, other organisms are finding it harder to concentrate the calcium carbonate from seawater to make their shells and skeletons. For an example, beginning in the mid-2000s, acidified water upwelling along the coast of America's Pacific Northwest prevented baby oysters from forming their first shell layer. This resulted in a collapse of the oyster industry in the region for numerous years running.

Third, hatching, maturation, courtship, egg laying, mating, molting, and many other bodily processes and life events of marine organisms often rely on pH signals to commence. Even subtle changes in pH can throw off the timing or cause the bypassing of these events. Males and females may mature or spawn at different times, resulting in mating failures, or predators may hatch before or after the prey they need.

## Why Should We Care about Acidification?

A study of corals on the Great Barrier Reef found that calcification declined by 14 percent over 20 years. While the corals themselves are not disappearing, their skeletons are thinning and becoming more brittle; these damaged corals are repairing more slowly, and young corals are often failing to grow.

Sea butterflies are in danger as well. One study found that as the aragonite saturation level decreased over 40 years, pteropods' shells became thinner and more porous. Pteropods are a primary food source for many species, including salmon, seabirds, and baleen whales, and changes affecting the prey will most certainly affect these predators.

## What Does This Have to Do with Jellyfish?

Laboratory experiments on the impact of ocean acidification on jellyfish have had mixed findings: some suggest that jellyfish are slightly affected, while others indicate no perceptible impact. The aspect of jellyfish physiology that appears to be most vulnerable to acidification is balance.

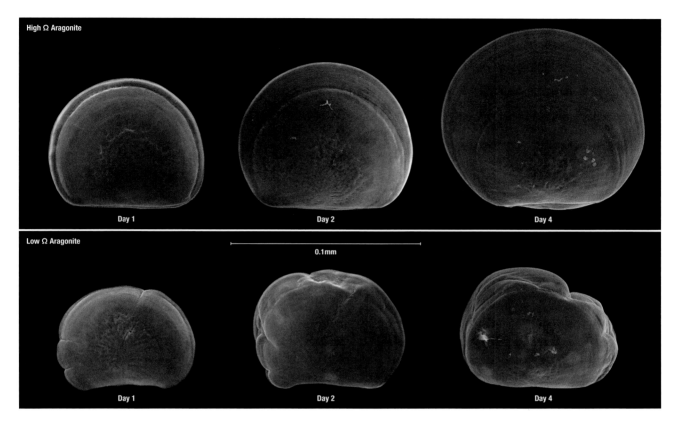

High Ω Aragonite

Day 1    Day 2    Day 4

Low Ω Aragonite

0.1mm

Day 1    Day 2    Day 4

**Above** Rising carbon dioxide ($CO_2$) levels in the atmosphere and ocean make it harder for organisms to build their shells. Newly hatched oysters are just tiny specks, but if they are unable to form their first shell layer, they will die. Those in the top row with normal rounded shells grew in favorable conditions, while those below with shell defects were grown in more acidic water.

**Right** Sea butterflies (pteropods) are a group of delightful, tiny free-swimming marine snails. In a process similar to osteoporosis, their shells become pitted and brittle, corroding away as seawater becomes more acidic. Researchers found that 53 percent of pteropods sampled along the west coast of America had severely dissolved shells, like the one figured in this electron microscope image.

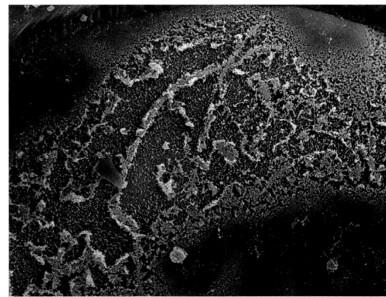

Many types of jellies have either one relatively large rock-like balance stone or numerous tiny fragments in each sense organ. Experiments simulating severe acidification have shown that these stones are affected, with the change causing the jellyfish to swim erratically. Even so, it seems clear that by the time jellyfish suffer any substantial ill effects of ocean acidification, other species will have long since suffered major repercussions.

# T R A S H

Plastic bottles at the beach, soda cans on the side of the road, plastic bags blown against fences: we seem to see trash everywhere nowadays. Anti-litter campaigns may be helpful, but there are aspects of trash pollution that most of us never think about—and these are changing our world.

ALL TRASH IS NOT EQUAL. Metals, glass, and electronic components often get recycled because there is a lucrative market for them. Paper too is frequently recycled, or it eventually biodegrades. But plastic is different. Most plastics are not recycled, do not biodegrade, and have no commercial value. They just persist. It has been said that every piece of plastic ever made still exists. We are burying our planet in plastic.

Many trash items end up in the ocean, whether they are blown, dumped, or carried there through waterways. Plastics are made to be durable, and indeed they are. On land they may take decades to centuries to break down into smaller pieces, depending on the amount of heat and ultraviolet light

to which they are subjected. Underwater they may never break down, for the heat and light necessary for the process do not penetrate into most of the ocean.

Even under the best of circumstances, most plastics only break into smaller and smaller pieces but never disappear. The sand on any beach, anywhere in the world, now contains tiny fragments of plastic as part of the sediment. These small pieces, along with tiny particles such as the abrasive plastic beads in skin cleansers, are known as microplastics.

Macroplastics, microplastics, and fragments of varying sizes are accumulating in the ocean at an alarming rate. Surface waters of the great ocean gyres (enormous, circling currents) have essentially become soups of floating or drifting

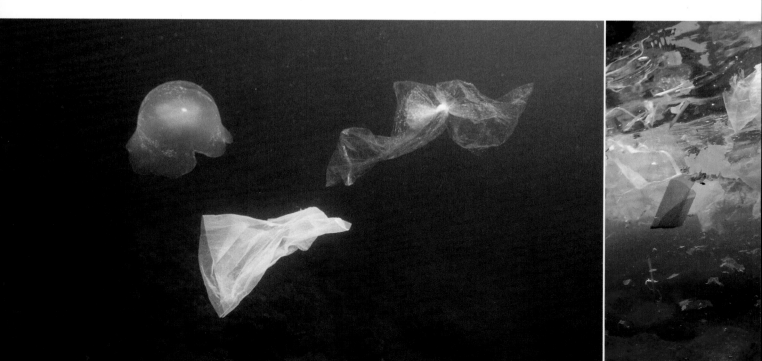

plastic. A study in the North Pacific gyre in 1998 found that plastics outweighed plankton six to one; within a decade that figure had risen to 46 to one.

Many scientists posit that we have entered a new geological era, basing this argument on such evidence as mass extinctions of animals and plants, rapid changes to climate, acidification of the ocean, and the proliferation of plastic. Plastic is spontaneously cementing with rocks and gravel and making a whole new type of rock, dubbed "plastiglomerate." Some scientists have proposed that the new era be called the Plasticene.

**Below left** Plastic shopping bags and cellophane wrappers look very similar to jellyfish, particularly for marine life like turtles, whose eyesight is poor and who are naive to the hazards of plastic. So too, dolphins may get plastic in their blowholes, perhaps mistaking it for jellyfish to play with.

**Below** Vast quantities of plastic are adrift in the sea, particularly in the middle of the world's largest oceans. These areas where floating plastic outweighs plankton are known as Great Ocean Garbage Patches. Astonishingly, floating plastic is just the tip of the iceberg; the majority is at the bottom of the sea or has been ingested.

## Why Should We Care about Trash?

Trash interferes with our aesthetic enjoyment of the world around us. More often than not, it interferes with the organisms we like and nurtures those we do not, such as cockroaches and rats.

Particularly in the ocean, small plastics and microplastics are attractive food items for many types of organisms, as they are brightly colored, easy to see, and easy to catch. Birds mistake floating cigarette lighters, straws, and bottle caps for fish. Fish mistake small beads and fragments for copepods or larvae. Even nonvisual creatures like clams ingest plastic. But they persist in the gut and give the animal a sense of satiety; many organisms starve to death with a gut full of undigested plastic. Moreover, the surfaces of plastics act as a magnet for toxic chemicals, accumulating these substances in concentrations many thousands of times heavier than they are in the surrounding seawater. Plastic therefore becomes a toxic bullet for the organisms that consume it—or the organisms that consume them, including us.

## What Does This Have to Do with Jellyfish?

Trash of any kind—particularly plastic—is a dream come true for jellyfish. Both flotsam (floating trash) and jetsam (sunken trash) provide new surfaces for jellyfish polyps to settle on, while flotsam also provides a convenient platform for dispersal. Floating items also serve as camouflage for jellyfish, protecting them from would-be predators. The brightly colored trash diverts attention away from jellies, or it may simply confuse a predator with too many choices.

Moreover, trash helps jellyfish win the survival game against other species. As microplastics are preferentially consumed by visual predators, more plankton is left for tactile feeders such as jellies, and as plastics sicken and kill their predators, the ecospace broadens for jellies.

# CHEMICAL POLLUTION

Chemical pollutants are toxic substances such as pesticides, heavy metals, dioxins, industrial residues, chemotherapy residues, and countless other harmful materials. There are, in fact, tens of thousands of potentially hazardous chemicals circulating around in the food we eat, the water we drink, the air we breathe, and the things we touch.

THESE CHEMICALS come from a variety of sources. Many help us, some hurt us. They may have benefits in small doses but be harmful in larger doses. In the environment, many of these chemicals can cause unforeseen complications. Pesticides, for example, help increase crop yields by annihilating unwanted insects, but they also kill indiscriminately, taking out the harmless and the helpful ones, such as bees, butterflies, and other pollinators; and they impact birds as well by eliminating or poisoning their food sources. Another example is the hormones that make it possible to raise livestock in vast numbers quickly, which do funny things to fish and other species when they wash into water bodies. Similarly, unsanitary overcrowding in cattle, poultry, and fish farms requires heavy use of antibiotics, and this is hastening the development of antimicrobial resistance, making antibiotics less effective for humans.

The vast majority of chemicals in use today have not been tested for their impact on human health. Moreover, the synergistic effects of these chemicals are almost completely unknown. Adjacent farmers using different pesticides may inadvertently combine them in toxic mixtures in the runoff puddles between their fields. A plate of salad or a bowl of soup may bring together chemical residues from different farms in combinations that were never meant to meet. We may even be creating Franken-chemicals in our own bodies, where residues stored in fatty deposits may come into contact with others, creating entirely new and untested combinations with completely unknown health effects.

## Why Should We Care about Chemical Pollution?

The rates of certain cancers are on the rise. The incidence of neurological disorders such as Parkinson's disease and developmental disorders such as autism is also on the rise. Children's bodies are maturing earlier than ever before. These changes have coincided with the increasing presence of manufactured chemicals in our daily lives, and yet we still wonder why such health problems arise. Untold numbers of chemicals are known to injure or kill cells, and some may promote cancer by increasing the rate of cellular turnover and therefore the rate of potential genetic errors that lead to the disease. Pesticides kill insects by disabling their neurological pathways. Hormones manipulate maturation pathways. Many types of maladies that threaten our health and well-being may be traceable to chemical exposure.

**Left** Jellyfish may flourish in the most polluted places, even where other organisms cannot. Without doubt this is due in part to jellies' brief lifespan and primitive organ systems. Their simple bodies are less likely to be affected, and if they are, their fleeting life soon ends the impact.

It is simply not possible in today's world to live free of environmental toxicants. Pesticide residues on fruits and vegetables cannot always be washed off. Hormones are in milk and in the tissues of meats. Heavy metals and other fat-soluble residues concentrate up the food chain. We store our foods in plastic long enough for it to leach toxic substances, and microwaving them in plastic accelerates that process. We eat foods from cans lined with chemicals. We breathe exhaust fumes from cars and industrial plumes, whether we see them or not. Lakes and bays are the ultimate repositories of unbelievable quantities of chemicals, and we simply have no way of knowing what is in the water we drink. When tested, bottled water often turns out to be more contaminated than tap water.

Our bodies may be exposed to a particular toxic chemical—or combination of chemicals—many times over the course of our lives, the exposure eventually manifesting in disease. It is rarely possible to pinpoint exactly which exposure might have led to a given case of illness.

## What Does This Have to Do with Jellyfish?

Jellyfish are among the few organisms that benefit from pollution. They are immune to the effects of many types of chemicals because their lifespan is too brief to allow cancer or other chronic diseases to develop. Moreover, their bodily structure precludes illness. Cancers that affect bones, brain cells, or the liver of fish or humans will not affect jellyfish, because they lack these organs.

While the downside to chemicals is minimized in jellies because of their biology, their ecology actually maximizes the overall ecological harm by chemicals, making it more difficult for populations of other species to recover from damage. In some cases, jellies are among the last species remaining after fish and other more complex species have succumbed to the toxic effects of chemicals.

# EUTROPHICATION

*Eutrophication* is the technical term for an aquatic ecosystem's response to nutrient overload. The nutrients are generally fertilizers, or the breakdown products of sewage, manure, and urine. Eutrophication, therefore, is too much fertilizer in the water, which often leads to depletion of oxygen. Whether this leads to hypoxia (low oxygen) or anoxia (no oxygen), the result is the same: a moonscape-like dead zone forms, where fish, crabs, oysters, and many other organisms cannot survive. These conditions occur in freshwater bodies such as lakes and rivers as well as coastal habitats like estuaries and bays.

A PULSE OF NUTRIENTS into a body of water can create unfavorable changes to the ecosystem. A common scenario occurs when a good strong rain washes accumulated nutrients from land into rivers and streams, and these enter shallow coastal bays and gulfs. The nutrients' outwash may include agricultural fertilizers, livestock waste, aquacultural and sewage effluent, pet and garden waste, and detergent phosphates.

These nutrients stimulate a bloom of phytoplankton, or plant plankton, which in turn provides unlimited food for zooplankton, or animal plankton. As these short-lived plankton die, their tiny bodies rain down through the water column by the thousands. Bacteria on the bottom decompose the plankton corpses, using up oxygen to drive their metabolism. The water stratifies into layers, with the deoxygenated bottom water sitting below a well-oxygenated and nutrient-rich upper layer. As the nutrients continue to support the plankton bloom and the decomposition cycle, creatures living on the bottom run out of oxygen. Those that can swim, scurry, or creep (such as crabs and sea slugs) flee, while those that cannot escape (clams, sponges, and sea anemones) simply die, adding to the decomposing biomass.

These eutrophic areas, called dead zones, now extend, in a perforated line, around the continents of the world. One by one, these perforations are joining. One day the continents of the world may be outlined by dead zones.

## Why Should We Care about Eutrophication?

The aquatic areas most often subjected to overnutrification—coastal seas and freshwater bodies—are the very ones we rely on the most. Coastal seas are the sites of most of our fishing and aquaculture. Freshwater bodies give us much of our drinking and irrigation water. As we go farther and farther afield to access wholesome fish stocks and clean water, these resources become more expensive. Moreover, in many areas it is no longer possible to look elsewhere, as those places are suffering the ravages of a growing population and urbanization too.

Dead zones have memory, meaning that once an area has gone anoxic or hypoxic, it develops a propensity to do so again. The lack of oxygen suffocates not only the obvious animals such as clams and crabs, but also less obvious species. Worms and microbes, for example, are the bioturbators (earthmovers) and the decomposers, and they keep the sediments healthy, just as their counterparts, earthworms and fungi, do in terrestrial environments. When they are killed off, the ecosystem may take a long time to resume normal function. Some dead zones are evergreen, lasting through the year, while others reestablish every year when storms subside and the water restratifies. In many areas the decline toward eutrophication is a one-way process, particularly as the human population continually rises and adds more poorly managed waste.

## DEAD ZONE FORMATION

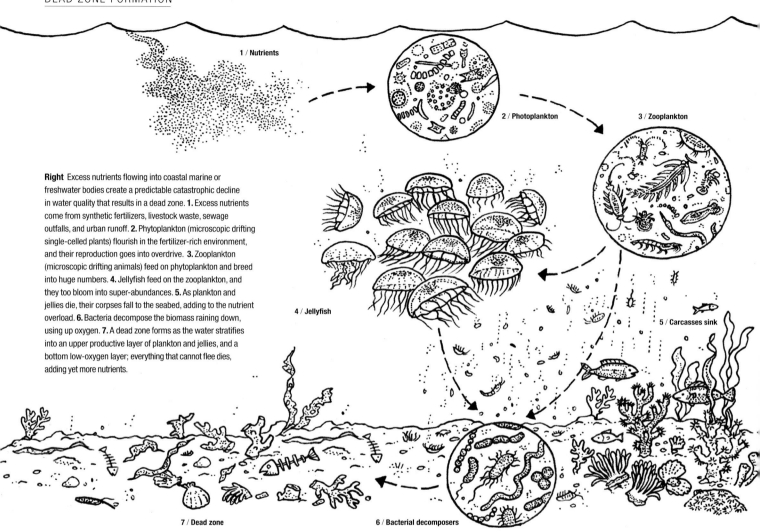

**1 / Nutrients**

**2 / Photoplankton**

**3 / Zooplankton**

**Right** Excess nutrients flowing into coastal marine or freshwater bodies create a predictable catastrophic decline in water quality that results in a dead zone. **1.** Excess nutrients come from synthetic fertilizers, livestock waste, sewage outfalls, and urban runoff. **2.** Phytoplankton (microscopic drifting single-celled plants) flourish in the fertilizer-rich environment, and their reproduction goes into overdrive. **3.** Zooplankton (microscopic drifting animals) feed on phytoplankton and breed into huge numbers. **4.** Jellyfish feed on the zooplankton, and they too bloom into super-abundances. **5.** As plankton and jellies die, their corpses fall to the seabed, adding to the nutrient overload. **6.** Bacteria decompose the biomass raining down, using up oxygen. **7.** A dead zone forms as the water stratifies into an upper productive layer of plankton and jellies, and a bottom low-oxygen layer; everything that cannot flee dies, adding yet more nutrients.

**4 / Jellyfish**

**5 / Carcasses sink**

**7 / Dead zone**

**6 / Bacterial decomposers**

### What Does This Have to Do with Jellyfish?

Jellyfish are well suited to areas impacted by eutrophication. They stay up in the water column above the dead zone, in the rich layer of uneaten plankton, feeding on a soup of small organisms. Jellyfish feed constantly and readily take advantage of the excess food. The high density of plankton often reduces visibility in these nutrient-flooded areas, but as tactile predators, jellyfish are not bothered by this. Finally, jellies' rate of oxygen consumption is incredibly slow compared to that of most other animals, and they are able to store oxygen in their gelatinous tissues, which allows them to descend into the oxygen-free waters from time to time with no ill effects.

In many areas, jellyfish have taken over the role of top predator above the dead zones to the extent that they exclude fish. With the impenetrable curtains of their deadly tentacles, these jellyfish rule the water column, and any prey or foe that ventures into the area is eaten, stung, or both.

# COASTAL CONSTRUCTION

As an environmental term, coastal construction encompasses the building of any artificial structure on or in coastal waters. It includes buildings, oil rigs, sea-based wind farms, ports and marinas, piers, aquaculture facilities, retaining walls and breakwaters, and underwater pipelines and cables. The list goes on; even shipwrecks and ocean dumping could be included.

ALL THIS CONSTRUCTION FAVORS the weedy species, or those that colonize fastest. More often than not, these "weeds" include introduced or invasive species. Such weedy species as dandelions and cockroaches owe their success to their fast growth rate, short life cycle, large number of offspring, and broad tolerance to a variety of conditions.

Weediness, therefore, is a lifestyle, not a taxonomic division. Like the common cold, weeds are always around, waiting for a teeny bit of ecospace to open up so they can stake their claim. An overturned rock, plowed soil, a felled tree, or a volcanic eruption can set the stage; weedy species just need a fresh place to plant some roots, make a nest, or lay their eggs.

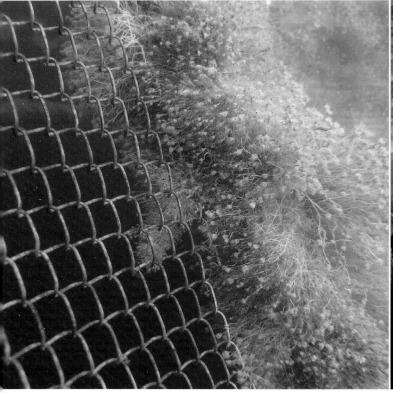

Among native species, a predictable ecological succession often covers surfaces: the initial colonizers appear, then a series of progressively larger, slower-growing species arrives, and eventually a climax community, or the stable ecosystem that we recognize as forest or coastal heathland or coral reef develops. Weeds also follow a succession, but it is less predictable and may become stable without reaching the native climax community stage.

Colonization occurs more rapidly in the sea than on land because of the density of "seeds"—in the marine sense this is often larvae of invertebrates. Most marine species have planktonic larvae, and many of these need a place to settle in order to continue their life cycle. In this soupy mix of larvae in highly productive coastal ecosystems, man-made structures offer the most opportunistic larvae a helping hand in the struggle for survival.

### Why Should We Care about Coastal Construction?
By adding structures, we permanently change the ecosystem dynamics of our coastal waters. These changes go far beyond aesthetics. The natural three-dimensional habitat (such as reefs, rocky outcrops, and canyons) that supports marine organisms becomes damaged. Dredges mow it down. Power plant intake pipes suck it out. Ports fill it in. All sorts of coastal activities pollute it into oblivion. Three-dimensional habitats are replaced with new habitat that often supports a different set of species, or none at all.

**Far left** Hydroid colonies rapidly overgrow man-made structures. This biofouling causes myriad problems for aquatic industries. Growth on aquaculture facilities creates drag on cages and stings fish stock. Proliferation on intake pipes impairs the flow to power plants and desalination plants. Fouling on marinas, ports, and ships can be hazardous and expensive.

**Left** Scyphozoan polyps prefer artificial surfaces, particularly on the undersides. Pontoons and docks often use plastic and polystyrene for floatation, and oil rigs and ports use metal and concrete. These provide millions of nooks and crannies, which are often teeming with feeding polyps (white) and those cloning baby jellyfish through strobilation (red).

Furthermore, these coastal structures often support the coming and going of ships, rigs, or pleasure craft that may transport invasive species, thus hastening their colonization. One need only examine a few research reports on introduced marine species to begin to see the potential for complications. A San Francisco Bay study in 1998, for instance, found that a new species has been introduced approximately every 14 weeks for the last 50 years, resulting in at least 234 exotic species with established populations, plus another 125 of uncertain origin. Another example is the port of Hobart in Tasmania, Australia: After decades of toxic industrial effluents and a highly invasive starfish, a biodiversity survey in 2011 found no native species.

### What Does This Have to Do with Jellyfish?
It is an ever-reliable rule of thumb that whatever is bad for fish is good for jellyfish; coastal construction, moreover, is really good for jellies. The proliferation of artificial structures associated with coastal industries such as shipping and aquaculture, along with structures involved in stabilizing shorelines, creates an enormous area of fresh surfaces to be settled. In scyphozoans in particular, polyps almost always suspend from the undersides of horizontal surfaces such as rocks and shells. Artificial structures such as piers, pontoons, aquaculture cages, seawalls, oil rigs, and pipelines have many nooks and crannies with large and small overhangs onto which polyps can stake out free space. In many places, these artificial overhangs substantially outnumber naturally occurring undersurfaces.

Laboratory experiments have demonstrated that jellyfish polyps prefer artificial substrates over natural ones. In particular, many studies have shown that when jellyfish larvae are given a choice of substrates, they frequently choose such plastics as polystyrene or polyethylene over natural surfaces like shells or wood.

# PUTTING IT ALL TOGETHER

While all the human-caused stressors discussed in this chapter help drive jellyfish blooms, a few are emerging as the main culprits: depletion of a certain type of fish, reduced oxygen in coastal waters, and warming. In the most troublesome jellyfish blooms at least one, usually more, of these factors is at play.

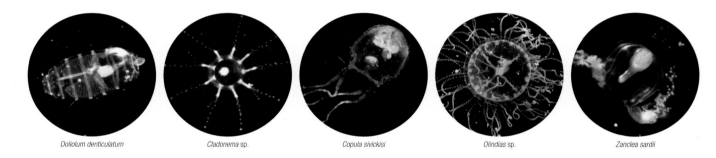

Doliolum denticulatum    Cladonema sp.    Copula sivickisi    Olindias sp.    Zanclea sardii

WHILE DEPLETION OF ANY FISH SPECIES destabilizes the ecosystem and can benefit jellies, the small pelagics, or forage fish—anchovies, sardines, pilchards, and menhaden—seem to be directly in balance with jellyfish. Numerous studies have come forth in recent years demonstrating a seesaw between the two: remove these fish and the jellies get out of control, or increase the jellies and the fish decline.

Oxygen also seems to play a particularly important role in jellyfish blooms. Coastal areas suffering lowered oxygen due to eutrophication are often dominated by jellyfish. Whether this is passive on the part of the jellies (they are one of the few creatures that can still live there) or active (they are in some way helping create the problem) is not yet clear.

When it comes to warming, jellyfish are in their element. Jellies just love warming. Springtime triggers them to bloom and summer is when they really flourish. Even subtle warming excites their metabolism and revs up everything they do: eating, growing, breeding. Warmer water, therefore, means earlier blooms, longer growing seasons, and larger infestations.

## The Other 90 Percent

The media, industry, and scientific communities have begun to pay attention to observable increases in jellyfish, but one important factor has been largely ignored: the 90 percent of jellyfish that are small, transparent, and inconspicuous. Most jellyfish measure about a centimeter (less than half an inch) across, some only a millimeter or two (less than the thickness of a dime). When we look over the side of a pier, or even peer out through a facemask underwater, these small species are almost impossible to see. They are visible only with proper lighting or if we drag a fine mesh net through the water (almost always guaranteed to produce amazement).

Surprisingly, only a single study exists on blooms of these small medusae. This survey revealed an astonishing pattern: the population underwent a complete turnover of species dominance and a five-fold increase in biomass in 20 years. This raises a question about similar effects in other areas where blooms of larger jellyfish are a problem: What is happening with the other, invisible 90 percent?

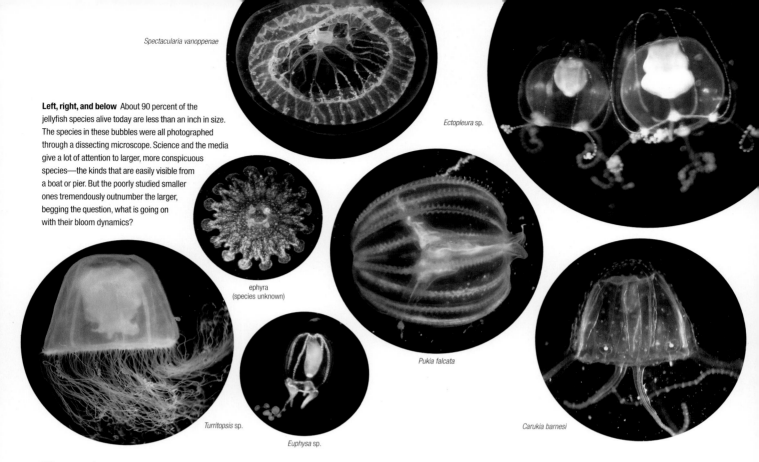

*Spectacularia vanoppenae*

*Ectopleura sp.*

**Left, right, and below** About 90 percent of the jellyfish species alive today are less than an inch in size. The species in these bubbles were all photographed through a dissecting microscope. Science and the media give a lot of attention to larger, more conspicuous species—the kinds that are easily visible from a boat or pier. But the poorly studied smaller ones tremendously outnumber the larger, begging the question, what is going on with their bloom dynamics?

ephyra
(species unknown)

*Pukia falcata*

*Turritopsis* sp.

*Euphysa* sp.

*Carukia barnesi*

## Time to Act

The Earth is like a patient who has had numerous tests with all the results indicating cancer: the time has come to act. No more data will change the certainty that there is an environmental problem poised to impact on humanity for a long time to come. With jellyfish, many argue that we have already passed a tipping point, and that as the environmental triggers of fish depletion, reduced oxygen, warming, coastal construction, pollution, and acidifying waters continue to increase, the problem will only get worse.

Some research groups and entrepreneurs have already begun to develop industrial eradication methods, while others are developing commercial uses for jellyfish. For an example of the former, in 2013 a developer created a device that is essentially a robotic blender, cruising through the water and chopping everything in its path. It is likely to cause more harm than good for many industries, but at least it is a start in thinking outside the box about ways to manage jellyfish blooms.

Commercial value may be found in the jelly that makes up a jellyfish body, which has many interesting properties, including elasticity, absorbency, and bounce. Many jellies also have fascinating biological features such as powerful toxins, bioluminescence, regenerative ability, cloning, and immortality. Some of these are already proving useful to us. For example, bioluminescence is being used to improve cancer surgery and track brain development. A research group has developed a method for using jellyfish as a low-calorie filler for muffins and cupcakes, while a commercial enterprise is using jellies as the thickening agent for caramel candies.

There are essentially two ways to manage the problem of jellyfish blooms: deal with the blooms or deal with the triggers. Commercial uses for jellies will provide the impetus for mass harvesting, which may give fish a fighting chance. However, if we do not address the underlying causes of the blooms—the human-caused stressors—reducing the jellyfish populations may just give the next pest in line a hand up.

# NOMURA'S GIANT JELLYFISH

Some of the most shocking reports about jellyfish in the last couple of decades have come from Japan. In 2000, a veritable sea monster awoke there. Prior to that, *Nemopilema nomurai*, or Nomura's Giant Jellyfish, was generally considered fairly rare and manageably small.

### Giant Problems

Now this species grows to the size of a refrigerator and blooms in uncountable numbers almost every year. It begins its life in the overpolluted and overfished waters off China. With elevated nutrients entering coastal seas from China's huge cities, hardly any fish left to eat the plankton, and ample coastal construction to give the polyps room to expand their colonies, *Nemopilema* flourishes. As the medusae grow, they drift northward on the warm currents toward Japan.

It has been credibly estimated that these giant jellies drift into the Japanese fishing grounds at a rate of a *half billion* per day. These throngs stifle the fishing industry. Fishermen are unable to draw in their catch due to the weight of the jellyfish and are forced to sever their nets. The jellies even capsized a trawler in 2009. These days, most fishermen do not even bother trying to go out when the jellies are around—it is just not worth it. In addition to damaging the fishing industry, *Nemopilema* has been blamed for at least nine fatal stings in China.

The Japanese government has gone to great efforts to manage this species, including attempts to develop industrial uses for it. Even though jellyfish are considered a delicacy in Japanese cuisine, the pest *Nemopilema* has a status similar to that of the rat in Western cultures.

**Scientific name** *Nemopilema nomurai*

**Phylogeny** PHYLUM Cnidaria / CLASS Scyphozoa / ORDER Rhizostomeae

**Notable anatomy** enormous, globular body with eight highly frilled oral arms bearing innumerable filaments

**Position in water column** epipelagic, in neritic zone

**Size** bell to nearly 2 m (6 ft. 7 in.) in diameter, weight up to about 200 kg (440 lbs.)

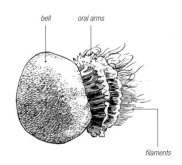

bell    oral arms

filaments

*Distribution*

# SEA WALNUT

THE NAME *Mnemiopsis leidyi* (*nee-mee-OP-sis LAY-dee-eye*) gives the shivers to the people who know of this ctenophore. It is one of the world's most unwanted pests. Native to the Western Atlantic, *Mnemiopsis* invaded the Black Sea via ballast water and completely devastated the ecosystem. In fact, within just a few years of its introduction, this single species of jellyfish multiplied so prolifically, it eventually comprised 95 percent of the Black Sea's biomass. *Mnemiopsis* then expanded into the Baltic and other seas of Europe, including the Mediterranean.

## Rapid Reproduction

The reasons for this sea walnut's success can be found in its weedy biology and ecology. It has special tentacles and food-capturing surfaces adapted for catching both large and small food particles. Feeding constantly, this species is able to eat more than 10 times its own weight each day, enabling it to double its own body volume in that short time. This extreme appetite and growth fuel its rapid reproduction. *Mnemiopsis* begins laying eggs within 13 days of its own birth, and by day 17 it lays up to 10,000 eggs daily. It does not even need to mate to produce young: *Mnemiopsis* is a self-fertilizing simultaneous hermaphrodite, meaning that each individual is both male and female at the same time and routinely fertilizes its own eggs.

## Bolinopsis

As impressive as *Mnemiopsis* is, species of *Bolinopsis*, which are its close relatives, are far superior in their capacity to do harm. *Bolinopsis* digests faster, grows faster, and reproduces faster. It is also more widespread, found from surface to depth, in every zone, pole to pole. *Bolinopsis* lacks the name recognition of *Mnemiopsis* but may ultimately prove to be a far worse pest.

---

**Scientific name** *Mnemiopsis leidyi*

**Phylogeny** PHYLUM Ctenophora / CLASS Tentaculata / ORDER Lobata

**Notable anatomy** transparent, very soft, gelatinous egg-shaped body with two large lobes, four long comb rows on the lobe sides, and four shorter comb rows in between

**Position in water column** epipelagic, in neritic zone

**Size** body to about 10 cm (4 in.) long

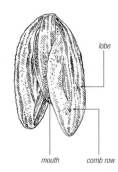

lobe

mouth          comb row

*Distribution*

# THE NOBEL JELLY

UPON ITS DISCOVERY IN 1902, no one imagined that the simple, elegant *Aequorea victoria* would become so famous. Its Frisbee-shaped body is completely clear except for the dozens of whitish lines radiating from near the center toward the edge, and it bears many cobweb-fine tentacles for catching its planktonic prey. It is one of the more beautiful of jellies and is a popular exhibition subject in public aquariums. But that is not why it is famous.

**Green Fluorescent Protein**

Under ultraviolet (UV) light or black light, nodes near the base of *Aequorea victoria*'s tentacles shine brilliant green. The molecule that makes this possible is known as green fluorescent protein, or GFP. Scientists have developed a way to join the GFP gene to other genes,

so that when the genes code for proteins, their processes and locations can be observed with UV light. Transbiological organisms that can be made to fluoresce have been created by grafting the GFP gene into plants, sheep, worms, mice, salamanders, fish, and other species—even cats.

GFP has been used in a variety of medical and biological applications, including tagging cancer cells, mapping the spread of HIV, and investigating how the brain translates sensory signals into motor output.

The scientists who discovered and developed GFP were honored with the 2008 Nobel Prize in Chemistry. As an ironic twist, however, intensive scientific harvesting of *Aequorea victoria*, beginning in the early 1960s and continuing for several decades, appears to have led to the depletion of its population. Whether it will recover is not yet clear.

**Scientific name** *Aequorea victoria*

**Phylogeny** PHYLUM Cnidaria / CLASS Hydrozoa / ORDER Leptothecata

**Notable anatomy** broad, shallow-domed, transparent body marked by up to 100 radiating canals and ringed with up to 150 fine tentacles

**Position in water column** epipelagic, in neritic zone

**Size** bell typically to about 8 cm (3 in.) in diameter

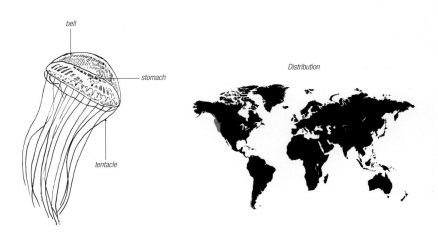

bell

stomach

tentacle

Distribution

# PSEUDO-IRUKANDJI

Long before *Malo kingi* was formally named and classified, it was dubbed "Pseudo-Irukandji," based on its similarity to the Common Irukandji (*Carukia barnesi*, page 154), the only Irukandji species then known. Its formal scientific name derives from two tragic stinging incidents: The species epithet *kingi* commemorates Robert King, an American tourist who was believed to have been killed by this species at the Great Barrier Reef. The genus *Malo* was named after Mark Longhurst (taking the first two letters of each of his names), a surfer who spent three days on life support after a sting by its sister species, *Malo maxima*, in Western Australia.

Drop for drop, the venom of *Malo kingi* and its close relatives may be the most toxic in the world, more virulent even than that of the Deadly Box Jellyfish (*Chironex fleckeri*, page 50). Just a small brush of the tentacle is all it takes to bring on Irukandji syndrome.

## Hypertension

The key to *Malo*'s hyper-lethality lies in a peculiar aspect of its venom. In addition to the severe pain, difficulty breathing, sweating, and other symptoms typically associated with Irukandji syndrome, *Malo* also causes a runaway increase in blood pressure. Hypertension as high as 280/180 has been documented, which can lead to stroke, pulmonary edema, and heart failure. *Malo kingi* has also been associated with priapism (prolonged erection) in male victims, and in that sense may prove pharmaceutically useful someday.

*Malo kingi* has a transparent, box-shaped body that is only two and a half centimeters (one inch) tall, and four tentacles, each up to 30 centimeters (almost a foot) long. It is found around tropical reefs and islands, and occasionally along the beach, in northeastern Australia. It eats small fish and shrimp, which it hunts with the aid of its keen vision.

**Scientific name** *Malo kingi*

**Phylogeny** PHYLUM Cnidaria / CLASS Cubozoa / ORDER Carybdeida

**Notable anatomy** small, rectangular, transparent body and four long tentacles

**Position in water column** middle to surface depths in shallow water, particularly over coral reefs

**Size** body to about 2.5 cm (1 in.) in height; tentacles to 30 cm (12 in.) long

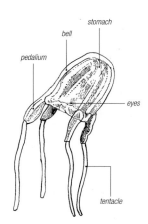

pedalium
bell
stomach
eyes
tentacle

Distribution

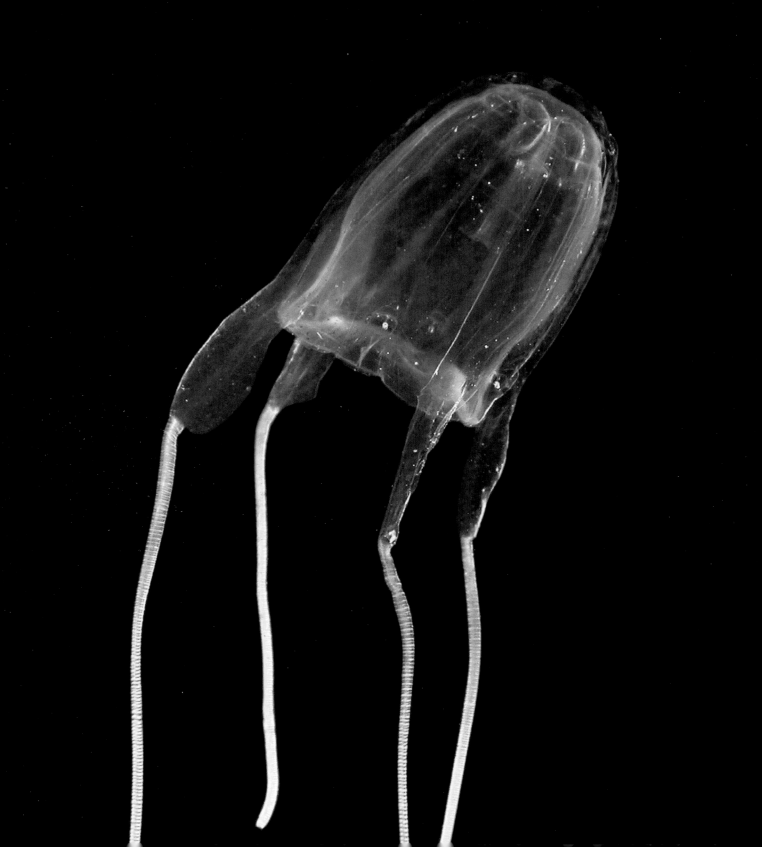

# SEA TOMATO

CRAMBIONE MASTIGOPHORA is native to Malaysia, where it is harvested for human consumption. After collection, the oral arms are lobbed off, and the bell is salted and dried in the sun. These dried jellyfish wafers, which store well, are then rehydrated in water, shredded into ribbons, and seasoned with sauce or dressing. But this jelly gets the common name Sea Tomato from its appearance, not its flavor: it is roughly the size, shape, and color of a large tomato.

### Australian Blooms

*Crambione* was first reported in Australia in the 1980s when huge numbers got sucked into the cooling intake pipes of a power plant, causing an emergency shutdown. Such blooms were rarely reported before 2000; now occurrences of epic abundance are reported almost every year. These blooms have to be seen to be believed: the jellies are packed in cheek by jowl for hundreds and hundreds of miles. One bloom lasted for 13 months. Another bloom stretched for almost 1,500 kilometers (nearly 1,000 miles).

The ecological effects of these blooms have only recently been studied. A primary component of the food consumed by the Sea Tomato is bivalve larvae, which could prove troublesome for the local pearl oyster industry. Moreover, the amount of plankton eaten by the growing bloom—and therefore not available to feed other species—and the strong pulse of decaying biomass as the bloom dies may cause major disruptions to normal ecosystem function.

Separate from its bloom dynamics and fisheries potential, *Crambione* is a beautiful species. Its nearly spherical, purplish-red body is smooth on top and bears eight feathery oral arms with many filaments.

**Scientific name** *Crambione mastigophora*

**Phylogeny** PHYLUM Cnidaria / CLASS Scyphozoa / ORDER Rhizostomeae

**Notable anatomy** round, blubbery body, deep red throughout, with eight short, feathery oral arms.

**Position in water column** epipelagic, in neritic zone; periodically beached in enormous numbers

**Size** bell up to about 25 cm (10 in.) in diameter

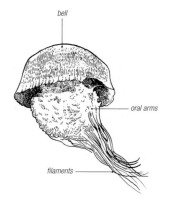

bell

oral arms

filaments

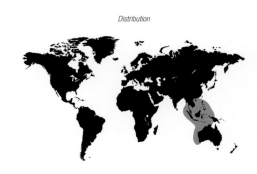

Distribution

# *OBELIA*

JUST ABOUT EVERY ZOOLOGY STUDENT learns of the *Obelia* jellyfish genus, as it provides the textbook example of the alternation of generations between medusa and polyp (hydroid) that is so common in jellyfish (see "Hydrozoan Life History," pages 66–67). *Obelia* is an imperfect representative, however, because it is one of the most aberrant of all medusae, lacking the bell-shaped body and flowing tentacles found in most species. Its tiny, flat, disk-shaped body (just a millimeter or two across, or about the thickness of a dime) is ringed by dozens of short, stiff tentacles resembling the rays of a sun in a child's illustration. The body contains five spots: the four reproductive organs and the mouth in the center.

## Hydroid and Medusa

Ecologically *Obelia* is a strange beast too, and its hydroids and medusae can be equally problematical, albeit in quite different ways.

The hydroids are bushy and tumbleweed-like, growing in great masses that may slow the rate of water flow through pipes or weigh down structures by creating drag. Contact with *Obelia* hydroids may cause urticating lesions for fish and other marine creatures with delicate skin.

Efforts to clean away fouling hydroids may result in the fragments simply seeding more colonies. Additionally, the mechanical action of breaking hydroids apart may in some cases stimulate them to grow more vigorously.

When stimulated by the right conditions, *Obelia* medusae bloom by the thousands and can present a stinging hazard for people or interfere with aquaculture. In fish gills, the medusae may damage the tissue sufficiently to promote bacterial or viral disease. Even though each medusa is tiny, its sting is powerfully painful, and swimming into a bloom of *Obelia* feels something like being blasted with hot sand.

**Scientific name** *Obelia* spp.

**Phylogeny** PHYLUM Cnidaria / CLASS Hydrozoa / ORDER Leptothecata

**Notable anatomy** small round, flat body with five spots; numerous short, stiff, radiating tentacles

**Position in water column** shallow coastal waters, particularly protected bays and harbors

**Size** bell 1–2 mm (0.04–0.08 in.) in diameter

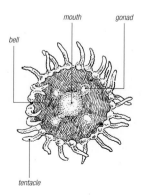

mouth   gonad

bell

tentacle

Distribution

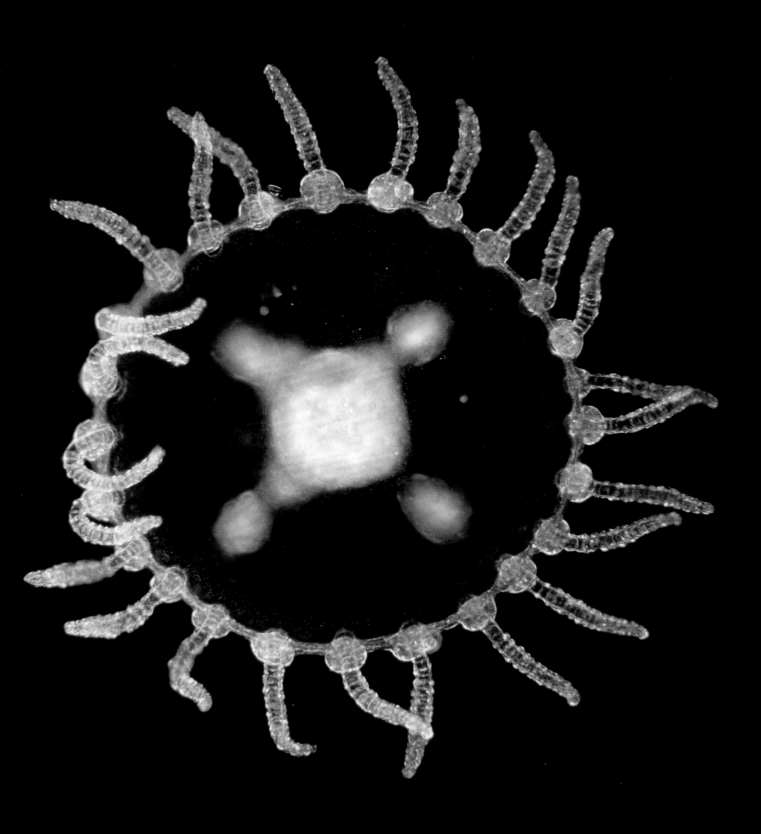

# LONG STINGY STRINGY THINGY

THE "LONG STINGY STRINGY THINGY" is, as one might imagine, long and stringy. And it stings. Numerous species fall into this informal group of siphonophores, or colonial jellyfish, all having similar structural features and ecology.

The most common of these species is *Rhizophysa filiformis*. Its colony members, known as persons, are situated along a main stem, which hangs from an air-filled bladder or float. The persons occur in repeating units of mouths, tentacles, reproductive organs, and so on. Many filaments and club-shaped appendages hanging from the central axis give these colonies their long, stringy appearance.

## Nasty Sting

Long stingy stringy thingies sting fiercely; the sting is comparable to that of their more famous relative the Portuguese Man-of-war. No fatalities are known, but the stings do cause the skin to peel away, as after a sunburn.

These creatures are rarely encountered by people because of their open ocean habitat. They can live from the surface to great depths, traveling between these regions by concentrating more gaseous molecules in their float to carry them upward, or burping off a bubble to descend back down. Upwelling events periodically bring deep-sea water masses toward coastal surface waters, and may transport these siphonophores into swimming areas.

Siphonophores that bear floats may give false sonar readings when they occur in high densities. The sonar waves bounce off the discontinuity in the water column produced by the air in the floats, creating the impression of a solid object. Masses of siphonophores have been reported as false bottoms and mistaken for fish schools. These days, one risk is that they may be interpreted as fish swim bladders, leading to overestimates of fishery stocks.

**Scientific name** *Rhizophysa* spp. and many others; the information below and the map are for *Rhizophysa* spp.

**Phylogeny** PHYLUM Cnidaria / CLASS Hydrozoa / ORDER Siphonophora / SUBORDER Cystonectae

**Notable anatomy** a spherical or oblong air-filled bladder, and a long stem bearing many filaments and clubs

**Position in water column** epipelagic, mesopelagic, bathypelagic

**Size** colonies typically from about 5 cm (2 in.) to nearly 1 m (3 ft. 3 in.) long

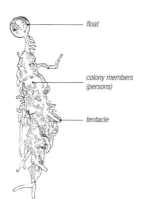

float

colony members (persons)

tentacle

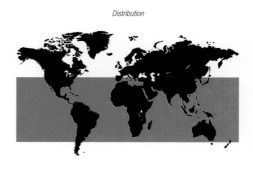

*Distribution*

# COMMON SALP

SALPS AND THEIR CLOSE KIN are probably the least familiar of the jellyfish groups. They do not sting, so they lack the notoriety of some jellies. But en masse they can strongly affect the ocean's ecological dynamic, stripping phytoplankton from the water, thus edging out slower-growing species such as krill and sea butterflies. Higher animals, including fish, whales, and seabirds, rely on these other species for food.

### Voracious Vegetarianism

*Thalia democratica*, like all salps, is a vegetarian. It eats huge quantities of phytoplankton, which enable it to grow up to ten percent of its body length per hour and to reproduce at a rate of two generations per day. When phytoplankton is abundant—following high-nutrient events such as upwelling or rain runoff—*Thalia* can exploit this banquet. Vast blooms are periodically recorded, especially off eastern continental shelf systems, including those of Australia and the United States.

### Salp-Irukandji Relations

Curiously, *Thalia* is strongly associated with Irukandji Jellyfish (*Carukia barnesi*, page 154) infestations in Australia and Thailand. The reason for this association is unclear. Neither eats the other: *Thalia* is a herbivore, while *Carukia* is a carnivore that eats tiny fish and prawns. Perhaps they just respond to the same environmental cues, or maybe the delicate Irukandji uses the tremendous gelatinous biomass of the salps as a kind of nonthreatening safety buffer, similar to the way fish find protection within a school.

### Carbon Sequestering

Salps naturally help sequester large quantities of carbon in the deep sea through the rapid sinking of their fecal pellets, thus contributing to reducing the carbon load in the ocean and the atmosphere. However, unsustainable numbers of salps would be required to keep up with our planet's carbon emissions.

**Scientific name** *Thalia democratica*

**Phylogeny** PHYLUM Chordata / SUBPHYLUM Tunicata / CLASS Thaliacea / ORDER Salpida

**Notable anatomy** small, barrel-shaped body with two gelatinous projections on one end

**Position in water column** epipelagic, in neritic zone

**Size** body 1–1.5 cm (0.4–0.6 in.) long

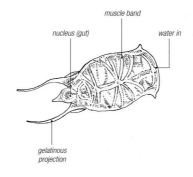

nucleus (gut) / muscle band / water in / gelatinous projection

*Distribution*

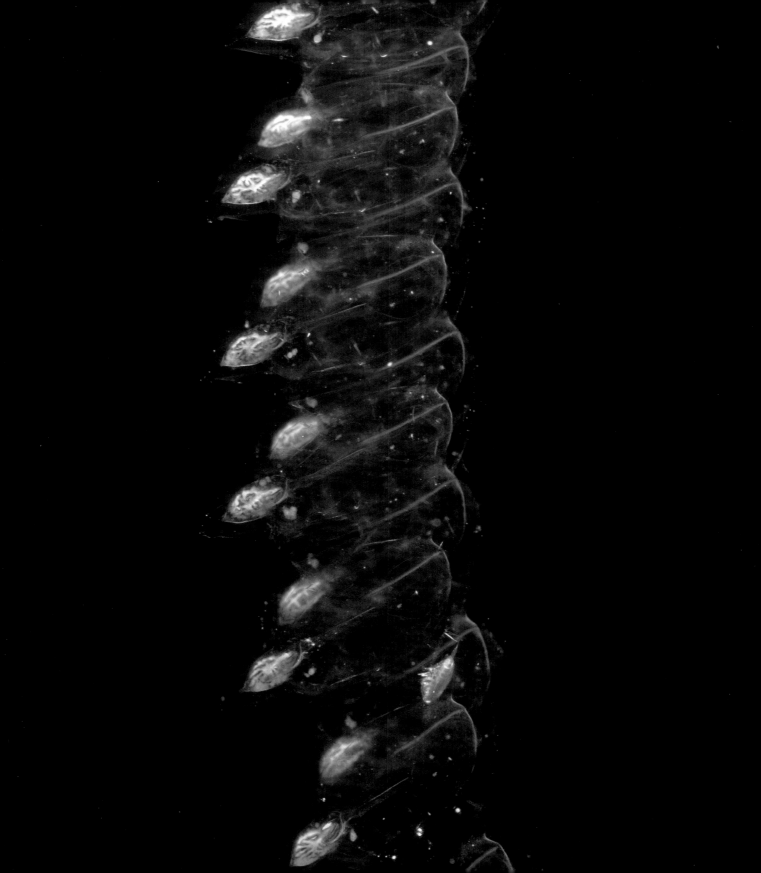

# BLUE BLUBBER

DESPITE ITS COMMON NAME, the Blue Blubber (*Catostylus mosaicus*) is not always blue. This species is quite common in the bays and harbors of southeastern Australia, particularly around Sydney, Melbourne, and Brisbane. In Melbourne, Blue Blubbers are blue. In Sydney, they are almost invariably brown. In Brisbane, they are often blue but are sometimes white with a dark blue margin, as they almost always are farther north. The brown color is explained by the presence of zooxanthellae, or symbiotic algae, similar to that found in corals, which masks the blue. The white is harder to explain, and may indicate that there are different species, as yet unrecognized.

### Blooming Blue Blubber

*Catostylus* has a hemispherical body with thick, heavy jelly. Eight conical oral arms, their surfaces cauliflower-like, protrude below. Blue Blubbers can get quite large and heavy,

and when mature can sting fiercely. The species eats plankton, but the algal symbionts supply most of the nutrition in those medusae that have them.

Some years, *Catostylus* blooms in such astronomical numbers that those describing it tend to run out of superlatives. Perhaps it is best known as the species that took on the US Navy. In 2006, the nuclear-powered aircraft carrier USS *Ronald Reagan* docked in the port of Brisbane, Australia, on its maiden voyage; the cooling pipes sucked in thousands of *Catostylus*, disabling the ship and necessitating an emergency evacuation.

Long-term data from the waters around Melbourne demonstrate a dramatic increase in *Catostylus* over the past several decades. The exact cause of this is not well understood but may be related to water quality or changes in local fish stocks or both.

**Scientific name** *Catostylus mosaicus*

**Phylogeny** PHYLUM Cnidaria / CLASS Scyphozoa / ORDER Rhizostomeae

**Notable anatomy** hemispherical body with eight conical, cauliflower-like oral arms protruding down

**Position in water column** shallow, coastal waters; generally found in bays and harbors

**Size** mature bell to about 30 cm (12 in.) in diameter

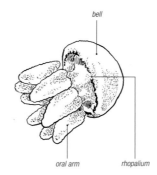

bell

oral arm

rhopalium

*Distribution*

# ATLANTIC SEA NETTLE

IN APPEARANCE, THE SEA NETTLES are the quintessential jellyfish, bearing the classic dome-shaped bell, four long, pleated oral arms, and 24 to 40 long, filamentous tentacles. Most are brightly colored, often with strikingly pigmented patterns of streaks or blotches, and are mesmerizing to watch in the water.

The Atlantic Sea Nettle (*Chrysaora quinquecirrha*) comes in two forms. The more common of the two is the white form, which is ghostly whitish throughout and has 24 tentacles; the so-called red form is white with a ring of 16 radiating red ovals on the bell, and has 40 tentacles. Probably, with modern scientific study, these two forms will prove to be different species.

Sea nettles prey on planktonic crustaceans and larvae by trolling their tentacles and oral arms through the water. Food is conveyed to the mouth between the bases of the arms. These jellies can also catch fish, which are digested externally on the oral arms.

## Declining Water Quality, Increasing Jellyfish

*Chrysaora quinquecirrha* can be the bane of existence to boaters, swimmers, and fishermen along the US eastern seaboard and the Gulf of Mexico. It harms countless people each year with its fiercely painful sting. It has been particularly bothersome ecologically in the Chesapeake Bay. The bay was once home to plentiful seafood and fish free for the taking. But everything changed when the bay's filter-feeding scallops and mussels were dredged out, which meant a slowing of the natural water-cleaning process. Urban runoff of nutrients stimulated algal blooms. As water quality declined, fish stocks crashed and jellyfish boomed, with the Atlantic Sea Nettle leading the charge.

Because of their beauty and strangeness, sea nettles have become a popular exhibit animal at public aquariums all around the world. Such displays give people a chance to see these notoriously dangerous creatures in a different, more enchanting light.

**Scientific name** *Chrysaora quinquecirrha*

**Phylogeny** PHYLUM Cnidaria / CLASS Scyphozoa / ORDER Semaeostomeae

**Notable anatomy** hemispherical body with four long, ruffled oral arms and 24 or 40 long filamentous tentacles; body all white, or white with 16 red radiating streaks

**Position in water column** shallow coastal waters, particularly bays and harbors

**Size** body to about 25 cm (10 in.) in diameter

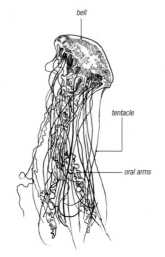

bell

tentacle

oral arms

Distribution

# GLOSSARY

**ABORAL** Opposite the mouth, or oral, side of an animal.

**ABYSSOPELAGIC ZONE** The portion of the ocean between 4,000 m and 6,000 m (13,000–20,000 ft.).

**APPENDICULARIAN** A tadpole-like pelagic tunicate of the class Appendicularia; also called a larvacean.

**BATHYPELAGIC ZONE** The portion of the ocean between 1,000 m and 4,000 m (3,330–13,000 ft.).

**BENTHIC** Of the bottom of a body of water, such as the seafloor.

**BIOLUMINESCENCE** The emission of light from a living being, such as a firefly or some jellyfish.

**CHITINOUS** Made of chitin, a hornlike or fingernail-like substance.

**CILIA** (singular, **CILIUM**) Hairlike or eyelash-like structures, such as the rows of projections that form the "combs" of the comb jellies. Ciliated means having cilia.

**CLADISTICS** A system of classifying relationships among organisms based on shared traits that have evolved from a common ancestor.

**CLONE** An individual produced asexually, by cloning, and genetically identical to its parent, rather than through sexual reproduction (from the union of egg and sperm).

**CNIDARIAN** A member of the phylum Cnidaria, a group of mostly marine invertebrates that includes corals, anemones, sea fans, and the jellyfish classes Scyphozoa, Cubozoa, Staurozoa, and Hydrozoa..

**COLLOBLAST** An adhesive, or sticky, cell found in ctenophores that is used in capturing food.

**COPEPOD** A tiny aquatic crustacean, often found in plankton, that is a major jellyfish food source.

**CTENOPHORE** A member of the phylum Ctenophora, the comb jellies, so named for their eight, comb-like rows of cilia. Examples are the sea gooseberries and the Venus's girdles.

**CUBOZOAN** A member of a class (Cubozoa) of highly venomous cnidarians known as box jellyfish. The medusae of these jellyfish are cubomedusae; their polyps are cubopolyps.

**DIOECIOUS** Having the male reproductive organs and the female reproductive organs in separate individuals. Compare *monoecious*.

**DOLIOLID** A small, barrel-shaped pelagic tunicate of the order Doliolida, distantly related to salps.

**ENDEMIC** Native to one particular place and not found naturally anywhere else.

**EPHYRA** A free-swimming larval medusa produced asexually from the polyp form of a jellyfish of the class Scyphozoa. See *strobilation*.

**EPIPELAGIC ZONE** The portion of the ocean into which sunlight penetrates and photosynthesis can occur, from the surface to about 200 m (660 ft.) deep.

**GANGLION** A mass of nerve tissue that may serve as a rudimentary nervous system.

**GENUS** (plural, **GENERA**) A taxonomic category between family and species comprising a group of closely related species. The first part of a species' scientific name is the genus; the Portuguese Man-of-war (*Physalia physalis*) and the Blue Bottle (*Physalia utriculus*), for example, share the genus name *Physalia*.

**GONAD** Reproductive gland; in the male it is the testes, which produce sperm, and in the female it is the ovary, which produces eggs.

**HADOPELAGIC ZONE** The portion of the ocean deeper than 6,000 m (20,000 ft.), extending to the seafloor.

**HERMAPHRODITE** An individual that has both male and female reproductive organs. It may be a simultaneous hermaphrodite, with the sexes occurring in the individual at the same time; or a sequential hermaphrodite, female first, then male (protogynous hermaphroditism), or male first, then female (protandrous hermaphroditism).

**HYDROID** The polyp form of a jellyfish of the class Hydrozoa.

**HYDROZOAN** A jellyfish of the cnidarian class Hydrozoa, which has three forms: those that occur mainly in a hydroid (polyp) form; those that have a dominant medusa form (called hydromedusae); and those of the order Siphonophorae (see *siphonophore*).

**INVERTEBRATE** An animal, such as a jellyfish, lacking a spinal column and an internal skeleton.

**IRUKANDJI** A term derived from the name of an Australian Aboriginal people used to refer to several species of box jellyfish that deliver highly venomous stings, as well as the syndrome, or physical constellation of symptoms, the stings cause in human victims.

**KERATINIZED** Converted into keratin, a horny material that occurs in human hair and fingernails.

**MEDUSA** (plural **MEDUSAE**) The free-floating form of a jellyfish of the subphylum Medusozoa, typically having a disk-shaped body (often called a bell) with oral arms and tentacles hanging below it.

**MEDUSOZOAN** A member of the Medusozoa, a subphylum of the class Cnidaria made up of the four classes Scyphozoa, Cubozoa, Staurozoa, and Hydrozoa.

**MESOPELAGIC ZONE** The portion of the ocean that lies from a depth of about 200 m to about 1,000 m (650–3,250 ft.), to which some sunlight penetrates but in a quantity insufficient to support photosynthesis.

**MONOECIOUS** Having the male and female reproductive organs in the same individual; hermaphroditic. Compare *dioecious*.

**NARCOMEDUSA** A jellyfish of the order Narcomedusae (class Hydrozoa), characterized by solid, rigid tentacles attached to the top of the bell rather than hanging from the underside.

**NECTOPHORE** A structure on a siphonophore that helps with locomotion by contracting and expanding; also called a swimming bell.

**NEMATOCYST** The stinging cells found on all members of the phlyum Cnidaria, including jellyfish (where they are usually found on the tentacles), used to stun prey and sting predators.

**NERITIC ZONE** The region of the ocean that lies over a continental shelf, from the low tide line to about 200 m (660 ft.) deep, generally within the reach of sunlight.

**NOTOCHORD** The defining feature of the phylum Chordata, a column of cells that is the precursor to the spinal column (vertebral column) in embryonic development in vertebrates and is also present in the larvae of some invertebrates, including salps.

**NUDIBRANCH** A shell-less marine mollusk, also called a "sea slug," that preys on cnidarians, including jellyfish, and incorporates its prey's stinging cells into its own tissues and uses them in its own defense.

**OCELLUS** (plural, **OCELLI**) A simple visual or light-sensing organ found in some jellyfish.

**ORAL ARM** A projection that hangs from the mouth area on the underside of a jellyfish, contains stinging cells, and is used in food capture. It usually occurs in fours, and is generally thicker and less strand-like than a tentacle.

**PELAGIC** Of the ocean's waters, referring to the entire water column from the topmost, epipelagic zone down to the hadopelagic depths.

**PH** A measurement of the concentration of hydrogen ions in a substance, such as water or soil, expressed on a scale of 0 to 14, with 7, the pH of pure water, being neutral. Numbers below 7 are acidic, and those greater than 7 are alkaline (or basic).

**PHARYNX** An anatomical passageway that in invertebrates generally refers to the portion of the digestive system connecting the mouth to the stomach.

# GLOSSARY

**PHOTOSYNTHESIS** The production of carbohydrates from carbon dioxide and hydrogen in cells that contain chlorophyll when they are exposed to sunlight.

**PHYLOGENY** The history of the evolution of species, genera, families, and higher groupings of organisms, tracing their lines of descent from common ancestors and configuring the relationships among them. A phylogenetic tree is a diagram showing, in the form of branches from a common trunk, these evolutionary relationships.

**PHYTOPLANKTON** The members of the plankton that are plants, including blue-green algae, diatoms, and dinoflagellates. These organisms carry out photosynthesis and are the primary element in the oceanic food chain. Although individually microscopic, they may bloom in large numbers and cover the water surface.

**PLANKTON** (adjective **PLANKTONIC**) Aquatic organisms, present in both freshwater and saltwater, that live in a passive, drifting state, with weak or no means of locomotion, and range from microscopic plants and animals to large medusan jellyfish and salps. Plankton includes algae, bacteria, crustaceans, mollusks, cnidarians, ctenophores, and larvae of various creatures. See *phytoplankton* and *zooplankton*.

**PLANULA LARVAE** A minute, free-swimming primary larval stage of many types of jellyfish that results from sexual reproduction (the union of eggs and sperm).

**PLATYCTENE** A creeping, flatworm-like type of ctenophore that dwells on the seafloor.

**PLEUSTON** Aquatic organisms that live in the zone of a body of water where the air and water meet.

**POLYP** The tiny, budlike life stage of many types of jellyfish that reproduces asexually by cloning.

**PYROSOME** A salp-like pelagic tunicate that occurs in colonies made up of individuals, called zooids, which together form large bioluminescent organisms.

**RADIAL SYMMETRY** Occurring in a circular, radiating form, similar to a daisy or a spoked wheel, that can be sliced into equivalent parts. Many jellyfish are tetraradial (divisible into four symmetrical parts), hexaradial (six-parted), or octoradial (eight-parted).

**RHIZOSTOME** A scyphozoan jellyfish belonging the order Rhizostomeae that lacks true tentacles and bears its stinging cells (nematocysts) on its oral arms; also called a blubber jelly.

**RHOPALIUM** (plural **RHOPALIA**) A sensory organ of medusozoan jellyfish that controls pulsations of the bell, balance, and the animal's visual or light-sensing capabilities.

**SALP** A jellyfish-like member of the class Tunicata that in its larval stage possesses a notochord. Salps occur in solitary and aggregate life stages, in the latter forming a chain composed of individuals (zooids or "persons") that serve different functions, such as food capture, digestion, and reproduction.

**SCYPHOZOAN** A jellyfish of the cnidarian class Scyphozoa, which includes the true jellyfish (sea nettles, moon jellies), the rhizostomes (blubber jellies), and the coronate jellies.

**SIPHONOPHORE** A jellyfish of the order Siphonophora (one of the major divisions of the cnidarian class Hydrozoa), which lacks a typical bell but has swimming bells and/or a float and employs tentacles for food capture. Portuguese Man-of-war is the most familiar siphonophore.

**SISTER GROUP** In cladistics, sister groups are those that are most closely related to one another; they are represented in a phylogenetic tree or similar diagram as equal, corresponding branches arising from the same ancestral origin.

**SPECIES** The basic unit in taxonomy representing a type or kind of organism. The members of a species are able to interbreed. Each species has a two-part scientific name (binomial) containing its genus and species designation (e.g., *Chironex fleckeri* for the Deadly Box Jelly). "Species" is often abbreviated, singular sp., plural spp.

**SPECIES CONCEPT** A paradigm, or theoretical framework, underlying the criteria scientists use for identifying and recognizing species. Different species concepts are based on the organisms' ability to interbreed (Biological Species Concept), their form or morphology (Morphological Species Xoncept), and their evolutionary relationships (Phylogenetic Species Concept).

**STATOCYST** In hydrozoan and scyphozoan jellyfish, a sac-like structure containing granules that stimulate different nerves as they shift about and provide signals to the animal regarding its balance and orientation.

**STATOLITH** In cubozoan jellyfish, a relatively large, stone-like structure composed of the mineral gypsum that is comparable in function to the granules in the statocysts of other jellyfish, helping the animal maintain balance and orientation. Also called a balance stone.

**STAUROMEDUSA** A trumpet-shaped cnidarian jellyfish of the class Staurozoa that lacks a free-swimming adult stage, living instead anchored to rocks or algae on the seafloor, with its usually eight tentacle-tipped arms flaring out in a star pattern. Also called a staurozoan.

**STOLON** In jellies, an elongated or bud-like extension from the body wall of a polyp, salp, or other form that develops into a new individual.

**STROBILATION** A form of cloning in scyphozoan jellyfish in which a polyp elongates and becomes segmented into a series of disks, each of which develops into an ephyra, a free-swimming larval medusa.

**SUBTIDAL ZONE** The shallow water of the ocean's neritic zone, close to shore but beyond the tidal zone and thus always underwater.

**SYMBIOSIS** A relationship in which two different species, each known as a symbiont, live in close association with one another. A mutually beneficial symbiotic relationship is that between upside-down jellies (genus *Cassiopea*) and single-celled algae inside them that provide nutrition through photosynthesis. A parasitic symbiotic relationship is the common association between jellies and hyperiid amphipods, small buglike crustaceans that burrow into and feed on the tissues of their host.

**SYSTEMATICS** The science of species classification (the taxonomic system) and species relationships (the phylogenetic system). See *phylogeny* and *taxonomy*.

**TAXON** (plural, **TAXA**) One of the units of taxonomic classification, such as species, genus, order, or phylum.

**TAXONOMY** The science of classifying organisms, including assigning them to a hierarchy of taxa that includes phylum, order, family, genus, and species, and assigning each a unique species name.

**TUNICATE** An organism, such as a salp or pyrosome, that belongs to the subphylum Tunicata, which is part of the phylum Chordata. In common with other chordates, tunicates have a notochord, though only in the larval stage.

**VERTEBRATE** An animal (such as a human being) with a spinal column.

**ZOOID** An individual member, also called a "person," of a colony-forming or aggregating organism such as a salp or pyrosome. Each zooid has a particular function within the colony; e.g., gonozooids and phorozooids serve in reproduction, while trophozooids feed and provide nutrition to the colony.

**ZOOPLANKTON** The portion of the plankton community made up of animal life, rather than plant life (see *phytoplankton*), including eggs and larvae of fish, clams, starfish, lobsters, and other non-planktonic animals, as well as permanently planktonic organisms, including single-celled protozoans, miniscule crustaceans such as copepods and the shrimp-like euphasiids (krill), sea snails, comb jellies, and medusan jellyfish.

# FURTHER RESOURCES

The following are some of the useful books, scientific journal articles,
websites, and apps currently available to those with an interest in jellyfish.

## GENERAL INTEREST BOOKS

Arai, M. N. *A Functional Biology of Scyphozoa*.
London: Chapman & Hall, 1997.

Bone, Q. *The Biology of Pelagic Tunicates*.
Oxford: Oxford University Press, 1998.

Gershwin, L. *Stung! On Jellyfish Blooms and the Future
of the Ocean*. Chicago: University of Chicago Press, 2013.

Sardet, C. *Plankton*. Chicago: University of Chicago Press, 2015.

Williamson, J. A., P. J. Fenner, J. W. Burnett, and J. Rifkin, eds.
*Venomous and Poisonous Marine Animals: A Medical and
Biological Handbook*. Sydney: NSW University Press, 1996.

## SCIENTIFIC JOURNAL ARTICLES

Boero, F., J. Bouillon, S. Piraino, and V. Schmid. "Diversity of
Hydroidomedusan Life Cycles: Ecological Implications and
Evolutionary Patterns." *Proceedings of the 6th International
Conference on Coelenterate Biology, 1995 July 16–21* (1997):
53–62.

Brotz, L., W. W. L. Cheung, K. Kleisner, E. Pakhomov, and
D. Pauly. "Increasing Jellyfish Populations: Trends in Large
Marine Ecosystems." *Hydrobiologia* 690 (2012): 3–20.

Condon, R. H., D. K. Steinberg, P. A. del Giorgio, T. C. Bouvier,
D. A. Bronk, W. M. Graham, and H. W. Ducklow. "Jellyfish
Blooms Result in a Major Microbial Respiratory Sink Of
Carbon in Marine Systems." *Proceedings of the National Academy
of Sciences* 108, no. 25 (2011): 10225–10230.

Duarte, C. M. and 19 other authors. "Is Global Ocean Sprawl a
Cause of Jellyfish Blooms?" *Frontiers in Ecology and the
Environment* 11, no. 2 (2012): 91–97.

Gershwin, L., W. Zeidler, and P. J. F. Davie. "Ctenophora of
Australia." *Memoirs of the Queensland Museum* 54, no. 3 (2010):
1–45.

Gershwin, L. and 10 other authors. "Biology and Ecology of
Irukandji Jellyfish (Cnidaria: Cubozoa)." *Advances in Marine
Biology* 66 (2013): 1–85.

Graham, W. M. and 14 other authors. "Linking Human Well-
being and Jellyfish: Ecosystem Services, Impacts, and Societal
Responses." *Frontiers in Ecology and the Environment* 12, no. 9
(2014): 515–523.

Mills, C. E. "Medusae, Siphonophores, and Ctenophores as
Planktivorous Predators in Changing Global Ecosystems."
*ICES Journal of Marine Science* 52 (1995): 575–581.

Mills, C. E. "Jellyfish Blooms: Are Populations Increasing
Globally in Response to Changing Ocean Conditions?"
*Hydrobiologia* 451 (2001): 55–68.

Purcell, J. E. "Jellyfish and Ctenophore Blooms Coincide with
Human Proliferations and Environmental Perturbations."
*Annual Review of Marine Science* 4 (2012): 209–235.

Totton, A. K. "Studies on 'Physalia Physalis' (L). Part 1.
Natural History and Morphology." *Discovery Reports* 30 (1960):
301–368, plates 7–25.

## FIELD GUIDES

Gershwin, L., M. Lewis, K. Gowlett-Holmes, and R. Kloser. *Pelagic Invertebrates of South-Eastern Australia: A Field Reference guide*. Hobart: CSIRO Marine and Atmospheric Research, 2013.

Book 2: The Medusae: https://publications.csiro.au/rpr/download?pid=csiro:EP1312312&dsid=DS2

Book 3: The Siphonophores: https://publications.csiro.au/rpr/download?pid=csiro:EP1312313&dsid=DS2

Book 4: The Ctenophores: https://publications.csiro.au/rpr/download?pid=csiro:EP1312314&dsid=DS2

Book 14: The Pelagic Tunicates: https://publications.csiro.au/rpr/download?pid=csiro:EP1312315&dsid=DS2

Kirkpatrick, P. A. and P. R. Pugh. *Siphonophores and Velellids: Keys and Notes For the Identification of the Species*. London: E.J. Brill/Dr. W. Backhuys, 1984.

Mapstone, G.M. *Siphonophora (Cnidaria: Hydrozoa) of Canadian Pacific Waters*. Ottawa: Canadian Science Publishing (NRC Research Press), 2009.

Wrobel, D. and Mills, C. *Pacific Coast Pelagic Invertebrates: A Guide to the Common Gelatinous Animals*. Monterey: Sea Challengers, 1998.

---

## USEFUL WEBSITES

**Australian Marine Stinger Advisory Services**
Information and downloads about jellyfish safety.
http://www.stingeradvisor.com/

**Claudia Mills Homepage**
Introductions and lists of valid names to the Ctenophora and Stauromedusae, tidbits on Hydromedusae, discussion of the bioluminescence of the jellyfish Aequorea, notes on Marine Conservation, and other information.
https://faculty.washington.edu/cemills/

**Freshwater Jellyfish**
Just about everything known about freshwater jellyfish.
http://freshwaterjellyfish.org/

**Jellies Zone**
Comprehensive coverage of all things jellyfish along the Pacific coast of North America.
http://jellieszone.com/

**Jelly Watch**
Record sightings of jellyfish and other marine life.
http://www.jellywatch.org/

**Plankton Chronicles**
Stunning imagery and film clips of all sorts of jellies and other zooplankton.
http://www.planktonchronicles.org/en.

**Siphonophores**
Just about everything you could want to know about siphonophores.
http://www.siphonophores.org/

**The Bioluminescence Web Page**
Great resource on all things bioluminescent.
http://biolum.eemb.ucsb.edu/

**The Jellyfish App**
Jellyfish identification, safety information, regional warnings and SMS alerts, ask an expert, community forum, upload photos, global coverage.
http://www.TheJellyfishApp.com/

**The Scyphozoan**
Comprehensive information on the Scyphozoa.
http://thescyphozoan.ucmerced.edu/

**Zooplankton of the San Diego Region**
Natural history information on jellies and other plankton, put together by the invertebrate staff at the Scripps Institution of Oceanography.
https://scripps.ucsd.edu/zooplanktonguide/

# INDEX

# INDEX

# PICTURE CREDITS

The Ivy Press would like to thank the following for permission to reproduce copyright material:

Alamy Stock Photo/blickwinkel: 106L; David Fleetham: 97, 161; Jeff Milisen: 43; Andrey Nekrasov: 101R; Sanamyan: 108; Visual&Written SL: 35; Roderick Paul Walker: 163; Waterframe: 81; Solvin Zankl: 145B.

Chuck Babbitt: 133

Jan Bielecki, Alexander K. Zaharoff, Nicole Y. Leung, Anders Garm, Todd H. Oakley CC-BY-SA: 123.

Biodiversity Heritage Library: 125.

Peter J. Bryant: 87.

By and by CC-BY-SA: 57R.

Corbis/Sonke Johnsen/Visuals Unlimited: 37; Latitude: 143; Andrey Nekrasov: 197.

© Ned DeLoach: 119.

Enviromet CC-BY-SA: 190L.

Gary Florin/Cabrillo Marine Aquarium: 115.

FLPA/Ingo Arndt/Minden Pictures: 94; Fred Bavendam/Minden Pictures: 137; Hiroya Minakuchi/Minden Pictures: 199; Norbert Wu/Minden Pictures: 17T, 157, 184, 203.

Larry Jon Friesen/Santa Barbara City College: 159.

Dr. Lisa-ann Gershwin: 29T, 29B, 33TL, 99L, 99R, 110R, 111L, 111R, 141, 146, 155, 192, 193, 201, 211.

Getty Images/The Asahi Shimbun: 179; China Photos: 178; Mark Conlin: 6; Ethan Daniels: 151T; David Doubilet/National Geographic: 51; George Grall/National Geographic: 213; Richard Herrmann/Visuals Unlimited: 17B, 105; Rand McMeins: 145T; Andrey Nekrasov: 129; Oxford Scientific: 148; Frederic Pacorel: 91; Alexander Semenov: 25R, 85, 104, 107, 109, 165; Lucia Terui: 195; Ullstein bild: 140.

© Karen Gowlett-Holmes: 110L.

© Image Quest Marine: 83.

© Rudy Kloser/Great Australian Bight Research Program 2015: 151B.

Library of Congress, Washington, D.C.: 102, 103.

Louis Mackay: 177.

© Eric Madeja/www.ericmadeja.com: 121.

© Alvaro Esteves Migotto: 89.

Takashi Murai/The New York Times Syndicate/Redux: 75.

© The Museum Board of South Australia 2016. Photographer: Dr. J. Gehling: 98

Naturepl.com/Aflo: 134; Ingo Arndt: 131; Brandon Cole: 149; Jurgen Freund: 54, 79, 136; Doug Perrine: 7L; Michael Pitts: 7R; David Shale: 169, 209; Paul D Stewart: 127; Visuals Unlimited: 24R; Solvin Zankl: 14.

NOAA, National Oceanic and Atmospheric Administration: 183B.

© OceanwideImages.com: 2.

© Chris Paulin/fishHook Publications: 171.

© 2016 Philippe & Guido Poppe—www.poppe-images.com: 207.

Denis Riek: 41, 117.

Science Photo Library/Alexander Semenov: 24L, 33B, 53; Wim Van Egmond: 45, 205.

Sea Life Mooloolaba, formerly UnderWater World: 17C.

Shutterstock/Vladimir Arndt: 186; Rich Carey: 185; Ethan Daniels: 147, 153; LauraD: 49; Red Mango: 5; Tororo Reaction: 9; Vilainecrevette: 1, 100, 101L.

Kåre Telnes, www.seawater.no: 77.

Image courtesy of George Waldbusser and Elizabeth Brunner, Oregon State University: 183T.

David Wrobel: 25L, 39, 47, 57L, 93, 106R, 142, 167, 173, 174, 190R.

# ACKNOWLEDGMENTS

With deepest and humblest gratitude I thank Mike Schaadt and the folks at the Cabrillo Marine Aquarium, for giving me my start; Freya Sommer and Dave Wrobel, formerly of Monterey Bay Aquarium, for sharing my jellyfish love affair; Chuck Galt, for opening the world of pelagic literature to me; Jeannie Bellemin, for totally getting the whole plankton thing; and Claudia Mills, Ron Larson, Dorothy Spangenberg, Dale Calder, Monty Graham, Mike Kingsford, Nando Boero, Bella Galil, Anthony Richardson, and the various other jellyfish researchers who have inspired me along the way. A very heart-felt thank you to Wolfgang Zeidler, Phil Alderslade, Peter Davie, Puk Scivyer, Rudy Kloser, and Tom and Tina McGlynn for believing in me, and to the many photographers whose work made this book possible.

The production team at the Ivy Press has been an absolute dream to work with: Tom Kitch, Kate Shanahan, James Lawrence, David Price-Goodfellow, Amy Hughes, Katie Greenwood, and Vivien Martineau … your work is so beautiful that my heart has skipped a lot of beats working with you guys, and only a few of these have been electrical. And with great warmth I thank Christie Henry and the gang at the University of Chicago Press.

For Patrick.

Lisa-ann Gershwin

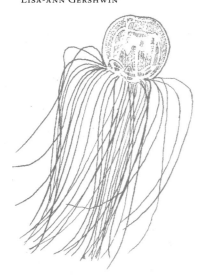

### She Dreamed She Danced with Jellyfish

Amidst the swirling currents ride
Strange jelly beings floating by
In reds and blues and brownish hues,
Far too many to classify.
Long Stingy Stringy Thingys,
An eyeball on a tassel,
With delicate pink tentacles belying
Its dreaded stinging hassle.
Bold Men-of-war armadas
Use nothing but the breeze,
Like sentinels from other worlds
An odyssey o'er the seas.
Wine-hued Giant Sea Nettles,
And silvery ones with purple stripes.
And ghostly Moons so splendid,
A zillion different types.
Twirling spheres and clapping lobes
Whose bodies catch the light.
Rows of flashing rainbows gleam,
Gifting me wild delight.
And Irukandji thimbles,
Wicked tentacles long and fine,
And as I reached to touch, the world
shifted from brine to mine.

Lisa-ann Gershwin & Phil Alderslade
*24 January 2016*

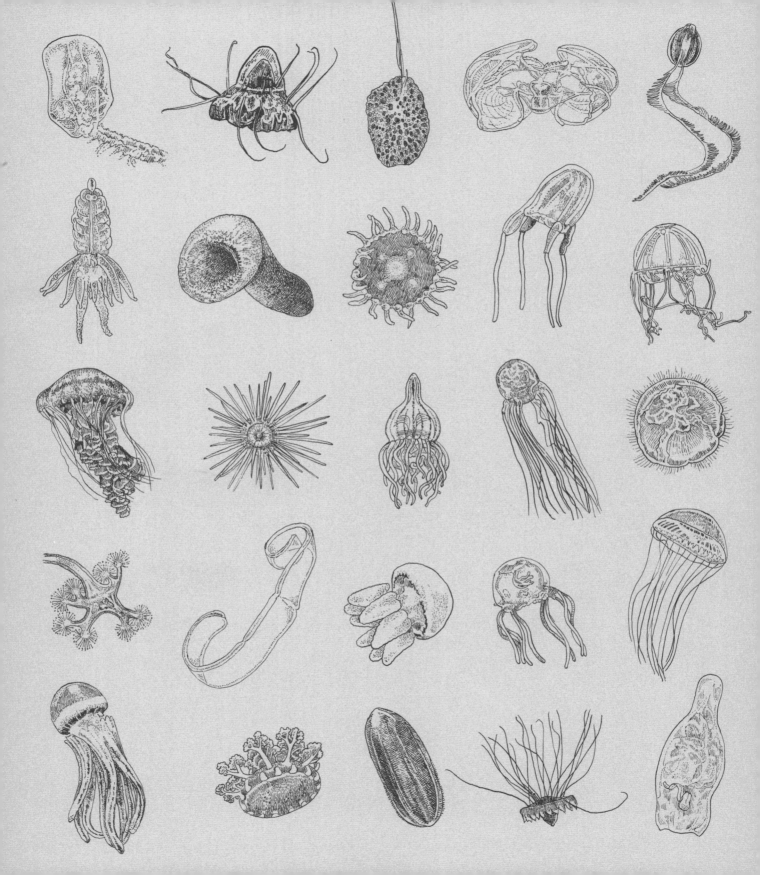